The Telecommunications Industry

THE DYNAMICS OF MARKET STRUCTURE

Gerald W. Brock

HARVARD UNIVERSITY PRESS
Cambridge, Massachusetts, and London, England 1981

Printed in the United States of America

Library of Congress Cataloging in Publication Data

Brock, Gerald W
The telecommunications industry.

(Harvard economic studies; v. 151)
Includes index.
1. Telecommunication—United States.
2. Telecommunication—Law and legislation—United States 3. Telecommunication policy—United States. 4. Telecommunication—Europe. I. Title.
II. Series.
HE7775.B68 384'.0973 80-25299
ISBN 0-674-87285-1

To my children
Jane, Sally, and David

Acknowledgments

Much of the research for this book was done at the University of Arizona. I am indebted to John Buehler, chairman of the Economics Department, for providing a teaching schedule conducive to research, and to my colleague John Drabicki for useful discussions on the formulation of the mathematical models. In addition, I benefited from a summer research grant and use of the university computer center. This work owes much to Richard Caves, who taught me industrial organization, guided my dissertation, and has continued to provide highly valued advice and encouragement. Sara Peterson typed the manuscript with great speed and accuracy.

I want to express special appreciation to my wife, Ruth, for her love, patience, and understanding throughout the time I have been working on this book.

Contents

Tables

Figures

The Telecommunications Industry

1 Introduction

An Industry in Transition

In 1978 American Telephone and Telegraph Company (AT&T) began a restructuring task that *Business Week* described as "the largest corporate reorganization in history."[1] The reorganization of the world's largest corporate employer reflected substantial changes in the telecommunications industry that it dominated. Spurred by a reduction in regulatory barriers to entry, many new companies entered the industry and challenged AT&T's monopoly status. Rapidly changing technology blurred traditional industry boundaries and brought AT&T into competition with established firms in other industries. AT&T responded to the changes in regulatory climate, technology, and industry structure with a barrage of new products, pricing policies, and political efforts designed to defend itself against the new competitors and to shape the telecommunications industry of the 1980s into a form favorable to the firm.

The most visible sign of change in telecommunications to the individual consumer was the transformation of the ordinary telephone from a standardized integral part of the telephone network to an electrical appliance with considerable variety. Before 1970 telephones were rented from the telephone company, were installed by a telephone company technician who visited the house and wired the telephone into the network, and were limited to a few basic styles and colors. In 1979 telephones could be purchased in many department stores and simply plugged in by the consumer. Styles and features proliferated as established telephone companies and new competitors battled to place additional phones in each home.

The business phone user encountered more substantial changes during the decade. New technology and new competition brought a wide variety of convenience features to business telephone systems. Electronic private branch exchanges (which switch telephone lines for a company's internal telephones) from the telephone companies and many competitors replaced the earlier electromechanical equipment. Reduced rates on private line long-distance service by new competitors and AT&T's responses to those rates increased the communications options of corporations. The changing rate levels and structures with the coming of competition also induced a tremendous increase in the complexity of determining least-cost communication paths and created a market for communications consultants and computer-controlled telephone routing plans to minimize costs.

Less visible, but potentially of even greater significance than the new entrants, was the blurring of the boundary between the telecommunications and the computer industries during the decade. Telephone switching systems became similar to large-scale computers, requiring manufacturers of telecommunications equipment to master computer technology. Communications became an indispensable part of distributed computer networks, requiring computer manufacturers to master communications technology. As the computer industry moved away from stand-alone computers toward networks of geographically separate but interconnected computers, and as the communications industry moved away from the pure provision of communications paths toward the provision of enhanced communications with computer technology, the boundary line between the two industries was blurred. By the end of the decade, AT&T had introduced a sophisticated "intelligent" computer terminal that it treated as a normal upgrade to its teletype terminals for providing communications, but that was perceived by others as an entry into the computer industry. IBM (through its partially owned subsidiary Satellite Business Systems) was planning a satellite communications system with rooftop antennas to enhance the transmission of computer data as well as provide voice communications services. Business writers regularly reported IBM as AT&T's chief competitor in spite of IBM's lack of any operational public communications system.[2]

Advances in technology also made feasible the rapid transmission of facsimile messages. Facsimile machines convert a written page into electronic signals, transmit those signals over telephone lines or other communications links, and reconvert the signals to a written page at the other end. While facsimile technology as such was not new, it had been a high-cost transmission method suitable only for specialized high-value communications such as news photos. Changes in technology during the 1970s greatly reduced the cost of facsimile service and induced several companies to enter the market for public facsimile service. IBM, Xerox, and the Post Office each planned sophisticated high-speed facsimile networks. While most companies viewed their facsimile efforts as an entry into the regulated communications market and applied for Federal Communications Commission authorization, the Post Office viewed its efforts as an extension of its traditional mail service. Rather than physically transporting letters from post office to post office, it would transport them electronically. Because electronic mail requires a combination of communications capability, computer capability, and traditional delivery services, the preliminary maneuvering for future position in electronic mail service brought interests from all three industries into competition. Potential competition was set up between a government monopoly corporation (the Post Office), a regulated monopolist (AT&T), and powerful unregulated firms with dominant positions in their home industries (Xerox and IBM).

Competition for the telecommunications market of the 1980s has proceeded on many fronts simultaneously. The firms have engaged in ordinary price and product competition, with each one attempting to convince the buyers that it can best fullfill their needs. They have competed for strategic technological advantages by pursuing sophisticated research and development designed to provide a decisive superiority over existing products. The firms have fought exhausting regulatory battles in multiyear proceedings with battalions of attorneys and support personnel. They have spent millions of dollars on lobbying and public relations efforts to obtain favorable legislation in Congress and state legislatures. They have made extensive use of the courts for antitrust suits, regulatory appeals, and other actions in the course of pursuing a competitive advantage.

Because much of the competitive activity is proceeding in public forums (courts, Congress, regulatory commissions) rather than only in the private market, the future direction of the telecommunications industry has become an important public policy issue. Much of the future of the industry will be determined by the decisions made in the various public proceedings on the industry. Consequently, it is important for individuals involved in the proceedings as well as the general public to understand the dynamics of the industry and the costs and benefits of alternative public policies. An understanding of the industry is also important for business firms that are heavy users of telecommunications in order to aid in rational planning for future services. The prediction of telecommunications directions is even more crucial for those firms currently or potentially involved directly in telecommunications services. The analysis in this book is designed to aid in understanding the telecommunications industry, to assist in formulating public policies toward the industry, and to help predict the future direction of the industry.

Preview of the Study

The primary purpose of this study is to clarify the causes of change in industry structures. It is well known that some industry structures change over time, but the causes of the change are not fully understood. Many economic writings assume that industry structures are relatively constant and treat changes as random, unpredictable occurrences.[3] It is the hypothesis of this study that structural changes occur in predictable ways as a result of economic and technological forces. The question of the predictability of structural change is not an esoteric, academic topic but has important public policy implications. If structural changes are purely random, policy formulations (antitrust, regulation, legislative changes) should be based on the currently observed industry conditions. If structural changes can be predicted, policy should be based on the expected conditions in the future. Thus it would be more appropriate to place constraints on an industry that currently has competition but is moving toward monopoly than on an industry that currently has little competition but is moving toward a more competitive structure.

The secondary purpose of the study is to examine the competitive uses of regulation and of systems effects. Regulation has many uses, including limitations on monopoly pricing, subsidization of some customers at the expense of others, and protection of existing industry participants from new competitors. All three have been examined extensively in previous literature. However, the normal treatment of regulation as a competitive device assumes a single monopolist or a unified cartel with the regulatory agency intentionally protecting the firms from outside competition.[4] Such a conception fit the telecommunications industry in the regulated period prior to 1956. More recent years have witnessed a relaxation of regulatory protection and the entry of new competitors. The recent period provides a good view of the ability of established firms to use regulatory procedures to their competitive advantage even when the regulatory policy is contrary to their wishes. The treatment of regulatory procedures as a part of the competitive process provides insight into the range of competitive activities available to firms and into the costs and benefits of imposing a regulatory framework on an industry.

Systems effects occur when the value of service offered by one firm is dependent on the number of customers served by that firm or other firms with which that firm exchanges business. Systems effects are prominent in communications and transportation. If railroads refused to connect their tracks and exchange cargos with each other, each one would have to extend its tracks all over the country in order to provide service to its customers. If telephone companies refused to connect their wires and exchange calls with each other, each would have to build a complete network in order to provide the communication desired by its customers. A nonconnected network reduces the value of service provided by all carriers, but makes the least reduction in the value of the largest carrier. Consequently interconnection conditions, permissions, and prices can be an important means of attaining and maintaining market power.

The tertiary purpose of this study is to provide an organized account of the development and current conditions in the telecommunications industry. Many works on telecommunications exist in the form of popular histories, government reports, scholarly articles, and regulatory filings. The specialist in telecommunications will find few facts here that could not have been located in other sources. However the fact that the data is public does not mean that it is easily available. Much of the information about the industry is contained in public filings with the Federal Communications Commission. Proceedings take on mammoth proportions with vast amounts of material filed in chronological order as it is received. Thus finding any given piece of information can be a nontrivial task. Data for this study comes from a great variety of sources including readily available published histories of the early days of telecommunications and less accessible filings in regulatory proceedings. The compilation of data within an economic framework is designed to provide readers with a convenient source of facts about the development of telecommunications and to assist in understanding the special economic conditions in that industry.

The underlying methodology for the study is drawn from the industrial organization branch of economics. That field provides theoretical and empirical insight into the relationships among the structure, conduct, and performance of an industry. Early industrial organization work focused on industry structure (primarily market share and the height of barriers to entry) as the primary determinant of industry performance (efficiency, progressiveness). Conduct was assumed to be determined by structure. More recent work in industrial organization, along with much of the business and legal literature, has given greater emphasis to the role of conduct as a determinant of both structure and performance.[5]

While drawing on the existing industrial organization framework, this study extends the theory to provide better insight into the process of industry dynamics. Neither the structuralist nor the behaviorist approach provides a comprehensive method of analyzing and predicting changes in industry structure. The structuralist approach assumes structure is relatively constant. The behaviorist approach assumes structure changes in response to conduct, but provides no explanation for the conduct except as a result of structure. The effect of exogenous technological progress on industry structure and the significance of varying opportunities to create market power for the development of industry structures have been inadequately explored.

The data for the study was selected to illuminate the role of industry conduct and technological progress in changing the structure of the telecommunications industry, along with additional material necessary to provide context and understanding of the events. The focus on changes in industry structure dictated a historical orientation rather than the more standard intensive examination of an industry at a particular time. Telecommunications has a long history, dating back to the beginning of commercial telegraph service in 1845. In the last 135 years, the industry has gone from a monopoly to competition to a cartel to a duopoly to a dominant firm structure to competition to a regulated monopoly to a regulated dominant firm. It thus provides ample opportunity for examining changes in structure, with increases and decreases in concentration under both free market and regulated conditions.

The last two sections of this chapter spell out the conceptual framework for analyzing regulation and systems effects. The following chapter develops a theory of changes in industry structure. It requires more economic and mathematical sophistication than the rest of the study, but a verbal summary is provided for readers who are disinclined to follow the technical reasoning. The empirical material begins with Chapter 3. Although the selection of the data and the analysis is based on established economic theory and the extensions in Chapter 2, the presentation is designed to be comprehensible to persons without formal economics training.

Chapter 3 highlights the problems of a cartel in taking collective action to protect its monopoly power. The original Morse patent was the basis for monopoly control of telegraph development, but the patent rights were held by a partnership without centralized control. Disagreements among the partners over the possibilities of outside competition and the best responses prevented effective joint action and led to competition in the industry. Monopoly control was restored by a new entrant (Western Union) through a careful program of mergers and exclusive right-of-way agreements with the railroads. Western Union was able to build barriers to entry to allow it to make high profits without losing control of the industry despite the expiration of the basic patents.

Chapter 4 emphasizes the role of a dramatic technological advance in breaking down monopoly power. The award of the telephone patent to Bell and the Bell Company's subsequent entry into telecommunications ended Western Union's dominance. Both companies possessed telephone-related patents potentially vulnerable to challenge from each other or outsiders. The brief, harsh competition between the two was settled by a cross-licensing agreement which restricted Western Union to telegraph and Bell to telephones. The agreement created a telecommunications duopoly. Each company gave up potential individual advantages in order to strengthen their combined defenses against outsiders. The expiration of the basic Bell patent and unfavorable (to Bell) rulings on other patents caused entry barriers to drop sharply in the 1890s. The continuation of a monopoly pricing structure and the reduction in barriers to entry brought in many new competitors and reduced Bell's control of telephones to approximately 50 percent of the total by 1907.

Chapter 5 is something of a digression from the main focus of the study. It summarizes early European telegraph and telephone development in order to provide some perspective on the events in the United States. The primary theme of the chapter is the interaction between public and private enterprise. In Britain, Sweden, and other countries, early telecommunications service was inaugurated by private companies, but by early in the twentieth century all major European countries had state monopolies of the service. The actions of the state companies to compete with the private companies were largely indistinguishable from the actions of nineteenth-century American companies intent on monopolizing an industry.

Chapter 6 traces the restoration of Bell's market control in the early twentieth century. Mergers were arranged with existing competitors. Public regulation was welcomed and used to prevent new entry into local service. Fundamental research and patent purchases were undertaken to protect the company from outside technology developments. The purchase of the DeForest vacuum tube patent provided Bell with needed technology for its own use and with a bargaining position with General Electric, RCA, and other companies that needed access to vacuum tubes. Patent licenses were used to establish boundaries between parties to the agreement and to provide joint protection against potential entrants to the respective industries. By the 1930s Bell had restored its dominant position in the industry.

Chapter 7 examines the use of regulation to maintain market power. The development of microwave technology for carrying television signals or large numbers of voice signals greatly reduced the technological barriers to entry into long-distance telecommunications. Microwave eliminated the need to secure a right-of-way to lay wires and allowed high-volume communication with a relay tower every thirty miles or so. The emergence of commercial television in the same period (late 1940s) and the need to distribute programs to local broadcasters provided a new source of demand for telecommunications. Without regulation, the reduction in barriers to entry at the same time as the arrival of a new source of demand would have brought many new entrants into the industry. Regulatory delays followed by an explicit policy against competition protected AT&T from the threat to its market control. A second threat in the form of a major Justice Department antitrust suit designed to break up the company (filed 1949) was broken with the aid of regulation. In a consent decree, AT&T was allowed to maintain its integrated structure and market control in return for freely licensing its patents and restricting itself to regulated activities.

The next three chapters examine the emergence of competition through changes in regulatory policies and technology. Chapter 8 focuses on long-distance services, Chapter 9 on terminal equipment, and Chapter 10 on changes in industry boundaries. All three chronicle the economic forces pushing the industry toward greater competition and the interaction of competitive forces and regulatory protection. Considerable attention is given to the competitive advantages that accrue to established firms from the regulatory processes even when the regulatory decisions are in favor of greater competition.

Chapter 11 summarizes recent congressional proposals for changes in the regulatory framework of telecommunications. Hearings on a telephone company sponsored bill to restore monopoly control over the industry have resulted in three new bills, all proposing a continuation of competition and gradual deregulation of the industry. A renewed Justice Department antitrust suit and numerous private suits also provide the potential for increased competition in the future. The study concludes with a summary of the conclusions which can be drawn from the history of the industry and the implications of those conclusions for policy directions.

The Theory of Regulation

Six different approaches to regulation can be identified: (1) the public interest approach, (2) the modified public interest approach, (3) the economic approach, (4) the status quo approach, (5) the political approach, and (6) the contract approach. Although the various views have contradictory assumptions and implications when treated as complete theories of regulation, they all provide insight into a certain aspect of regulation. This section synthesizes the various approaches into a conceptual framework for understanding the regulatory issues in telecommunications.[6]

The public interest theory is the fundamental basis for regulation in the law and in early economic writings. It assumes that regulated industries have special characteristics that prevent effective competition. Regulation is a substitute for market competition and protects the consumer from the arbitrary exercise of monopoly power by the producer.

The modified public interest approach assumes that regulation is designed to benefit the public interest by protecting consumers against monopoly power, but that it sometimes fails to do so because of administrative shortcomings. The problems of regulation are a result of remediable managerial and procedural difficulties: quality of staff, likelihood of staff being employed by the regulated industries, unequal resources available to agency and regulated firm, cumbersome procedures, excessive reliance on precedent, and so on. In this view, regulation is the proper response to technological limitations on effective competition, but reforms are needed in the administration of regulation to ensure that it operates in the consumer interest rather than only the producer interest.

George Stigler's economic theory of regulation dispenses with the assumption that regulation is assigned to industries with special technological problems that prevent effective market competition.[7] The demand for regulation comes from firms that can use regulation to enhance collusion, erect barriers to entry, restrict competition from substitute products, or obtain direct government subsidies, instead of from consumers who want regulation to protect themselves from a monopolist. The supply of regulation comes from political jurisdictions that are willing to meet the requests of the firms in return for political favors. Regulation is a device for redistributing income from consumers to producers. It is granted by the politicians because the small and generally hidden losses borne by the consumers do not produce adverse political consequences, while the large benefits gained by the firms result in their support in future elections.

The status quo theory of Donald Dewey, Bruce Owen, and Ronald Braeutigam[8] assumes that the long delays in regulatory proceedings and careful attention to past history and precedent are not failures of regulation but methods of fulfilling its primary purpose. The purpose of regulation is to protect the status quo and prevent parties from incurring the sudden capital losses (especially of intangible capital, such as market share or the opportunity to receive service below its true cost) that are intrinsic to the market system. The formal regulatory proceedings, such as extensive hearings on the allowable rate of return, are elaborate charades which have no real substance in themselves but help to preserve the existing prices and patterns of service. Regulation provides benefits to both consumers and producers by reducing risk and expediting planning. It imposes costs on any party that would have benefited from market-imposed changes from the status quo. As in Stigler's economic theory, the losing parties may be unaware of their losses due to regulation because of uncertainty over what pattern of prices and services would have existed in the absence of regulation.

John Baldwin's political theory of regulation views regulation as a method of arbitrating the disputes of producers and consumers.[9] The parties seek arbitration from the government in some form, such as requesting laws to benefit themselves at the expense of others. If the government intervened directly in the dispute, the losers would attempt to defeat the responsible parties at the next election. By creating an independent regulatory agency that is free from direct political control, the government can fulfill the arbitration function demanded of it while insulating itself from adverse political consequences.

Victor Goldberg's contract approach assumes that regulation is created to administer a long-term flexible contract between the consumers and producers.[10] The regulatory agency acts as an agency for the consumers in bargaining with the producers. A regulated industry is not intrinsically different from an unregulated industry which uses contracts that require interpretation and modification over time. In both cases, the producers and consumers are limited in their alternatives and settle their differences through bargaining (including potential use of lawsuits to interpret rights under the contract) rather than by choosing new suppliers.

An additional concept relevant to understanding regulation is Albert Hirschman's analysis of the roles of "exit" and "voice" in controlling organizations.[11] Exit is the market response—a dissatisfied customer chooses a different supplier the next time. Voice is the political response—a dissatisfied person attempts to change the unsatisfactory organization rather than leaving it. Most organizations are controlled by a combination of voice and exit. A competitive business firm will respond to customer complaints (voice) because they provide an early warning signal for potential sales reductions (exit). A state government normally controlled by the political process (voice) may change its policies in response to exit of dissatisfied customers desired by the state (such as large employers).

All firms are subject to external controls through the market or the political process or both. A market-controlled firm is limited in its action by competition from actual or potential rivals. If entry is easy, a firm may be market controlled even with no existing competitors, because potential competitors will enter if the existing firm makes it profitable for them to do so. Similarly a politically controlled firm may be limited by established regulations or the potential for direct control if its actions do not meet a particular standard. There is some tradeoff between market control and political control. A firm with significant competitors will be less likely to have political control imposed against its wishes than a monopoly.

From a dominant firm's point of view, total market freedom is impossible under the political conditions of the United States in this century. If it is able to eliminate actual and potential competitors (no market control or "exit" control), it can expect political control to be imposed through the existing laws (antitrust action) or new laws. Thus a firm that wishes to dominate its market must seek an explicit form of political control (such as a regulatory agency) which will allow the market dominance in exchange for certain powers given up to the regulators. The ideal solution from the firm's point of view would be to obtain nominal regulation, which imposed no significant restrictions on its conduct but protected it from other forms of political control and from market competition. The greater the restrictions imposed by the regulatory agency, the less likely the firm will be to support the regulation.

There is no evidence that regulation has been imposed over the determined opposition of the firms involved, though the firms may oppose the particular form chosen in favor of another version. Similarly there is no evidence that regulation has been imposed over widespread consumer opposition. Rather it appears that regulation is only imposed when both consumers and producers favor it, at least in the beginning. To gain the political support of all parties for passage of the necessary legislation, the new regulation must not make any party worse off than it was prior to the regulation. The legislation can gain the support of all parties if it protects their current positions against expectations of possible losses. Thus the essence of regulatory legislation is the establishment of property rights in the status quo for all parties. An additional aspect of regulation is the substitution of voice for exit. Both firms and consumers delegate authority to the regulatory commission to settle disputes subject to a variety of procedural safeguards, such as formal hearings. The parties give up some rights to market advantages in exchange for a formal mechanism, including an arbitrator, to bargain about changes in the status quo.

The establishment of property rights in the status quo and the transfer of departures from the status quo from the realm of competition to the realm of bargaining and arbitration cause the basic features of the industry prior to regulation to continue during the regulated period. Thus the price structure existing prior to regulation is generally maintained during the regulated period except for cost based changes. Preregulation discriminatory pricing features (higher business telephone rates than residential, value of service freight rates) are generally maintained under regulation with the description changed from profit maximizing monopoly price discrimination to socially desirable subsidies. The preregulation market structure is continued under regulation by placing limitations on either entry or exit. The preregulation service patterns are maintained under regulation by restricting service discontinuances and new kinds of service.

If regulation establishes property rights in the status quo, then all changes should be negotiated by the parties or imposed only after due legal process. Thus two commonly observed aspects of regulation, acceptance of agreements negotiated among the parties without independent examination and elaborate quasi-judicial proceedings prior to contested decisions, would both be expected. The parties are entitled to modify the contract by unanimous agreement, but disputed changes can only be implemented with careful regard to the interests of all sides. Parties who stand to lose from changes in the status quo must be given a fair hearing and sometimes compensated for their losses. With property rights, the status quo is the default position and any party wishing to move from there must prove his case. Thus prices can be raised for proper cause, service can be discontinued with a proper showing, and new entry can occur with proof that it is in the public interest by most regulatory statutes. However, all of those actions require considerable effort on the part of the party initiating the change, while little or no showing of benefit is necessary to maintain the status quo.

When the parties find their interests in harmony, or when the parties are able to work out private compromises of their conflicting interests, the regulatory agency has little direct effect. When the parties come to irreconcilable differences, the regulatory agency acts as a combination arbitrator and court. As arbitrator, it attempts to promote discussion and compromise for a satisfactory solution to all parties. As court, it holds formal hearings and makes definitive interpretations of the regulatory contract (subject to further review by appellate courts). It seldom if ever makes policy initiatives on its own but sets policy directions through its response to individual conflicts in the same way that courts do.

The conception of regulation as the establishment of property rights in the status quo can lead to the results predicted by any of the six theories of regulation. Regulation is in the public interest if monopoly power would be increasing without it, or if the public interest includes slowing the rate of change under market conditions. If the conditions are met for regulation to be in the public interest, but problems arise in achieving proper representation and administration, then the modified public interest theory comes into effect. For example, the commission may accept a privately negotiated agreement believing that it represents a compromise of all parties, while some affected parties were left out of the negotiations.

The predictions of Stigler's economic approach are fulfilled when the market trend would be toward less monopoly power without regulation. In that case, the regulatory protection of property rights in the status quo amounts to protection of existing firms to the detriment of consumers. The assumption that regulation preserves property rights in the status quo is very similar to the assumptions of the status quo theory, but somewhat less rigid. Regulation tends to maintain the status quo when compared to marketplace changes, but there is room for extensive changes with the consent of the parties. The quasi-judicial regulatory proceedings perform a real function of interpreting the rights of the parties.

Aspects of Baldwin's political approach can be seen in the separation of the arbitration process from the direct political process. Treating the arbitration of disputes as a matter for independent regulatory agencies rather than one for the elected leaders reduces the dangers to elected officials from disaffected losers. The emphasis on due process and compromise also creates less opportunity for political repercussions than do the administrative decisions of an executive agency. This approach is also consistent with Goldberg's contract theory. Regulation is viewed as a long-term, flexible contract with constraints on the freedom of the parties to negotiate a contract providing a different distribution of property rights than exists at the time regulation is imposed.

This view of regulation suggests that the effects of regulation increase over time. No change in the existing activities of the firms is imposed by the beginning of regulation. As conditions change, only negotiated changes or changes approved after a hearing can be implemented. Thus the primary effect of regulation is to slow down change. This will benefit parties that would have lost ground from the changes and hurt parties that would have gained from the changes. Potential entrants are made worse off by regulation because their normal function is to reduce the profits of existing firms. They must either compensate the firms for their losses, or prove to the regulatory agency that their entry will not have any significant effect on existing firms. Either is difficult and reduces the incentive to enter. Customers in an industry that would have become more competitive under market conditions will lose from the regulation, while the firms in that industry will gain.

Technological progress will generally be slowed by regulation. Progress will occur at a pace determined by the existing firms, but the ability of new firms to implement changes will be reduced or eliminated. For example, imagine what would have happened if the stagecoach lines had been placed under regulation prior to the building of the railroads. The regulatory agency probably would have blocked the expansion of the railroads in competition with stagecoach lines because the railroads imposed losses on the stagecoach operators. The stagecoach operators would have been free to adopt the technological innovation themselves if they chose to. If stagecoach operators were unable or unwilling to build railroads, railroads would not have been built. It is likely that such a system would have produced much slower railroad development than actually occurred.

Regulation works best in a time of technological stability. The reduction in risk and increase in ability to plan that come from protection against rapid changes may outweigh the disadvantages of regulation in an industry with little opportunity for technological change. But regulation is likely to slow or eliminate the benefits that could come from a major technological change. Large costs can be imposed merely by not adopting the best available technology. Many of these costs will be unknown. If stagecoach regulators had forbidden new railroad companies, the stagecoach customers would have been unaware of what benefits they could have gained from railroads. So long as the operators gave good and improving service and experimented with innovations such as railroads, it would have appeared that the stagecoach regulation was maintaining an orderly and effective transportation system.

Regulation can be used as an effective competitive weapon. Regulation provides the greatest benefits to a firm that faces a decline in its market power from other existing firms, new entrants, or antitrust action. Three sources of protection are available. First, the firm can draw on its formal rights in the regulation as enforced by the agency and courts to prevent usurpation of its power. Second, the firm can drawn on the procedures of the regulatory system to its advantage even if the final decision is likely to go against it. Extensive delays can be created by requiring formal hearings. Large costs can be imposed on other firms with proportionately greater costs on the smaller firms. While the requirement for formal procedures to implement contested changes in the status quo does not absolutely prevent the changes, it greatly increases the barriers to entry even if the agency is inclined to grant the change. Third, the firm can use its subjection to the regulatory agency as a defense against antitrust attempts to reduce its market power.

Systems Effects

A system is a set of complementary products which must be used together in order to provide value. That definition takes in a wide variety of products: records and record players, automobiles and gasoline, computer central processing units and disk drives, telephone sets and central office equipment, to name a few. The systems that are of primary competitive interest are those in which at least one firm sells a complete system. In that case, a variety of competitive uses of the system interrelationships can be made. Systems can be divided into public systems and private systems. A public system is one in which the individual user purchases or rents some capacity in the system rather than the entire system. Common examples include the telephone network and the railroad network. A private system is purchased in its entirety by the final user. The fundamental competitive opportunities from public and private systems are the same, but they are often perceived differently because of the difference in ownership.

The competitive opportunities inherent in private systems are illustrated by the experience of the computer industry. A computer system consists of various pieces of equipment (such as a central processing unit, tape drives, disk drives, and printers) which satisfy precise interface specifications for transferring data among the pieces. Before 1969 most computer equipment was sold as systems rather than as individual units. Users did not purchase, for example, an IBM disk drive to use with a Burroughs central processing unit (CPU) because the devices were incompatible. Barriers to entry were much lower into peripheral devices (tape drives, disk drives) than into complete systems. Consequently many companies began offering "plug-compatible" replacements for IBM peripherals after 1969. The fact that the new companies were competing against complete systems with only the ability to replace partial systems left them at a severe competitive disadvantage. IBM switched profit from the peripherals to the central processing unit and made various changes to the interface specifications and the leasing plans to discourage the invasion of plug-compatible peripheral companies. The competitive responses were all based on using the market power arising from barriers to entry into one component of the system (central processing units) to limit entry into other components.[12] The development of "plug-compatible central processing units" after 1975 greatly reduced the range of competitive opportunities because no part of the system was protected from entry.

The most straightforward method of combating the plug-compatible peripherals makers would have been to prohibit the connection of IBM CPUs with non-IBM peripherals. That was the goal of the various competitive actions taken. However that route was foreclosed because it would have constituted a tie-in sale in violation of the Clayton Act. IBM used such a prohibition prior to 1936 in requiring all users of its punch card machines to use only IBM supplied cards.[13] The antitrust laws restrict the freedom of private system suppliers to use systems effects to competitive advantage.

Two separate competitive advantages accrue to the systems supplier if it can tie the system components together. First, it can accept its monopoly profits on the systems components of its choosing rather than only on those with high barriers to entry. By using the interrelationship among the system components, the manufacturer can earn greater total profits than it could on the individual components. For example, if entry is difficult into CPUs and easy into disk drives, the nonsystems approach would be to take all monopoly profit on the CPUs and sell the disk drives at a competitive price. But if those who value the CPUs the most also use the most disk drives, greater total profits can be obtained from selling the CPUs at a competitive price and placing a high price on the disk drives.

The second advantage is a possible increase in total monopoly power. Total barriers to entry for a system may be more than the sum of the barriers to entry in each of its components. Greater capital and managerial requirements exist for entry into a complete system than into individual components of a system. The inventor of an improved component is more likely to become an active entrant if that component can be marketed alone than if it can only be sold as a small part of a complete system. To increase total market power through systems control, some method is necessary to prevent single-component manufacturers from entering in all the components and thus allowing a system to be constructed. The more single-component manufacturers there are, the less risk is involved if another single-component manufacturer enters the business. Thus in the computer industry, the existence of plug-compatible peripherals makers eased the way for the entry of plug-compatible CPU makers, and the entry of plug-compatible CPU makers increased the viability of the plug-compatible peripherals makers.

The telephone system is an example of a public system. Users do not purchase the entire system but only the right to use a part of the system. An intermediary (the telephone company) is interposed between the equipment supplier and the final user. Competitive systems effects arise from restrictions on interconnections among suppliers of service. A prohibition of interconnection is equivalent to a prohibition of mixing equipment types in the private system. However, because the prohibition comes from the telephone company rather than the equipment manufacturer, it is a refusal to deal instead of a tie-in agreement. Refusal to deal by a dominant firm is illegal if it is a component of an attempt to monopolize, but a greater range of defenses are available for refusal to deal than for a tying contract.[14] As in the private system, interconnection restrictions may increase the firm's choice of pricing strategies or increase total monopoly power by increasing barriers to entry. Systems effects provide no competitive advantage if there are no barriers to entry in any of the components of the system.

Economies of scale at the local distribution level provide some barriers to entry into telephone service. Each user typically has a dedicated access line to the nearest telephone switching office as an indivisible capital requirement for access to the system. For most residential subscribers and some business subscribers, one line has the capacity to carry all of their telephone calls. If two nonconnected telephone systems existed, the subscriber who wanted access to both would need two separate lines resulting in underutilized capital. For the heavy users of telephones, multiple lines are necessary even with a single system in order to handle their volume of calls. In that case, separate lines into separate networks would not necessarily result in underutilized capital. Similar economies of scale exist in public telegraph service. The telegraph distribution requirements consist of a telegraph line and office in each town. A single minimal office will more than handle the traffic of a small town, while multiple offices will be necessary in large cities. Thus multiple noninterconnected networks would be more viable for communication limited to large cities than for communication distributed to small towns as well.

The economies of scale in final distribution cause the most efficient method of communication to be either a single supplier or multiple interconnected suppliers. The private incentives to connect multiple systems depend on the sizes and locations of the various suppliers. Competitors of equal size have a strong incentive to interconnect. Interconnection increases the value of the service offered by each company because it can provide communication with more people than without interconnection. If any one company chooses not to interconnect, it suffers greater loss than its competitors. Interconnection rights can consequently be used as a means of disciplining a cartel.

If existing companies are of unequal size, interconnection provides maximum efficiency but benefits the smaller members more than the larger. Thus it is likely to be withheld if the larger company is attempting to monopolize the market. If instead of geographical separation, the smaller company's customers are also served by the larger company, the larger company has no incentive to provide interconnection. It gains no enhanced value from interconnection, because it could serve all the customers by itself that it could serve with interconnection. Thus it is unlikely to provide connecting privileges except under legal constraint.

In general, the value of access to a telephone system increases with the number of subscribers connected to the system. This gives an advantage to the largest system and makes competitive entry on a small scale difficult or impossible in the absence of interconnection. However, two qualifications that limit the competitive advantages of a larger system should be noted. First, the real variable of interest is not the total number of people connected but the number with whom the subscriber wishes to communicate. People with whom the subscriber does not wish to communicate are a neutral factor at best and a negative factor at worst. Some telephone users go to considerable lengths (unlisted numbers, a secretary to screen incoming calls) to limit their effective telephone network to a small segment of the total subscribers connected. Because most telephone users place the majority of their calls to a relatively small number of people, they may value a small telephone network that is limited to a particular neighborhood or other community of interest as highly as a much broader network.

The second qualification is that significant costs are incurred in achieving compatibility among all parts of a universal telephone network. A new innovation may be uneconomic for implementation in a large general purpose network but worthwhile for a separate specialized network. In such a case, the user is compensated for the limited scope of the network by the advantages gained from the noncompatible innovation.

2 The Theory of Dynamic Industry Structures

The traditional industrial organization assumption is that industry structure is relatively constant and is the primary determinant of conduct and performance. More recent writings have put increased emphasis on the role of conduct in shaping structure. Caves and Porter have developed a theory of mixed structural and created barriers to entry in which barriers to entry are created by the investment activities of the firm. Spence has examined the significance of strategic investments in capacity and in growth paths for the development of industry structure. Williamson has emphasized the role of varying firm responses to exceptional opportunities in the emergence of dominant firms.[1] The increased analysis of the role of strategic conduct in developing industry structure is consistent with the presuppositions made by European industrial organization writers, business writers, and antitrust law. European industrial organization has emphasized a greater role for conduct than the traditional United States analysis. Business writers generally stress the role of managerial decisions over the role of industry structure in determining firm performance. Antitrust law prohibits "monopolizing" and other actions rather than the existence of certain market structures.[2]

Three models can be used to extend previous theories of the influence of conduct on structure in order to develop a more complete framework for analyzing structural changes. The first one examines the effect of changes in opportunities to invest in barriers to entry on the final industry structure. The firm attempts to maximize the discounted present value of profits over an infinite time horizon. It can build barriers to entry by current expenditures. The higher the stock of barriers to entry, the higher it can set future prices without attracting entry. If price is set above the limit price determined by the competitors' costs and the stock of barriers to entry, competitors enter the market and reduce the dominant firm's market share. Entry is not instantaneous but occurs over time at a rate determined by the profitability of entry. A very high price attracts rapid entry; a price close to the limit price attracts slow entry. Both market share and barriers to entry are forms of intangible capital for the firm. Both are purchased through a reduction in current profits and both yield higher future profits.

The model shows that the height of barriers to entry created by the dominant firm will rise with increases in the responsiveness of new competitors to profitable entry opportunities and with the effectiveness of expenditures on creating barriers to entry. The height of barriers to entry will fall with increases in the firm's discount rate and with increases in the rate of depreciation of barriers to entry. The price and the market share of the dominant firm will move in the same direction as the height of barriers to entry in response to changes in the exogenous variables. Consequently an increase in the cost of barriers to entry or an increase in the rate of depreciation of barrier capital will cause the industry to become more competitive through the voluntary actions of the dominant firm. The increased costs of creating barriers will result in lower barriers to entry, lower prices, and higher competitive market share.

While the first model focuses on a single industry in isolation from the rest of the economy, the second model examines the impact of technological progress in a second industry on market power in the first industry. The technology of the second industry could be used to build the product of the first industry (electronic components to build watches, for example), but it is too costly for initial feasibility. Technological progress occurs more rapidly in the second industry than in the first industry, allowing a firm in the second industry to become a potential entrant without gaining access to the established technology of the first industry. From the point of view of the first industry, the technological progress is exogenous and threatening. It is not developed by or under the control of the industry and is better understood by the potential entrants than by the established firms. Such technological developments can break down carefully established barriers to entry and make a sudden transformation in the industry structure.

If both industries are monopolized and there is no collusion, the industry 2 monopolist will enter industry 1 on a small scale even while it has higher costs because of the profitable opportunities left by the monopoly pricing in industry 1. The new entrant will then continue to expand and become the dominant firm unless the original monopolist gains access to the new technology at reasonable cost. However, the most likely case is collusion. The fact that competition reduces total profits and the firms are initially in separate industries provides an opportunity for profitable, legal collusion. Industry 2 can sell its technology to industry 1 for prices equal or greater than the profits it would make by entry and agree not to enter. The monopolist of industry 1 would then remain a monopolist and would switch to the new technology when it became profitable to do so. It would have lower profits than without the threat from industry 2, but higher than if actual entry occurred. Although the economic effect is an agreement to divide markets and avoid competition, the legal status is a patent license or technology exchange between noncompetitive firms in different industries. It is unlikely to receive antitrust scrutiny. Another profitable opportunity for the monopolist is to seek regulation to prevent socially wasteful production through the entrant's initial higher costs. Because the monopolist's position continually declines under market conditions, a regulatory reduction in the rate of change is to the monopolist's advantage.

The third model considers a reduction in market power from a change in industry boundaries. No explicit new entry takes place. Instead, technological progress occurs in a related product. Initially the price of the second product is so much higher than the price of the first product that it is only used for specialized purposes and has no competitive effect on the first product. This was the case, for example, with long-distance telephone service in the 1880s compared to telegraph service. As the price of the second product drops relative to the first, the two come into closer competition and become differentiated products in the same industry. If the price of the second continues to drop relative to the first, the two products may move into separate markets again with the first product retaining only those uses for which the second is unsuitable.

If the firms do not recognize their interdependence, the second industry will gradually overtake the first. The second will appear to be an expanding industry and the first a dying industry. If the industries have a dominant firm, they will recognize their interdependence as the prices converge. The firm in industry 1 will perceive the firm in industry 2 as a new entrant and a threat to its market control. If industry 1 lacks the ability to enter the second product, it has little bargaining power and must accept the reduction in its market control. If entry is possible into product 2, the threat of entry can be traded for concessions from the firm in industry 2 in a negotiated settlement.

All three models assume that firms attempt to maximize their long-run profits. Although that is a traditional economic assumption, its validity can be questioned. The separation of ownership and control in large corporations removes the automatic link between profits and personal benefits to the managers. This has led many economists to advocate the substitution of a managerial utility function, including a broad range of goals (profits, growth, staff, working conditions, and so on) for the profit maximization assumption. The existence of regulation adds weight to the arguments against profit maximization, because profits are at least nominally controlled by the regulatory agency. While the challenges to the profit maximization hypothesis have considerable merit, the managerial utility function approach also has drawbacks. The multiplicity of possible goals greatly reduces the predictive content of a theory based on managerial utility functions and comes close to the vacuous assertion that managers do whatever they do because it fulfills their goals.

The underlying assumption of the models of this chapter and the empirical material that follows is that firms maximize their long-run ability to earn profits. When incorporated into a mathematical model, this assumption is indistinguishable from the assumption of profit maximization. However, it is more general than ordinary profit maximization and is meant to apply to unregulated firms, regulated firms, and publicly owned firms. Maximizing the long-run ability to earn profits is equivalent to developing maximum market power. Market power is beneficial to managers even if they cannot share any profits from that power (as in the case of a public firm). With market power, the firm may accept its "profits" as actual accounting profits or in many other forms, including slack and the pursuit of "pet projects" which could not be cost justified under competition. The assumption focuses attention on externally observable conduct, such as prices charged, new products introduced, and legislation sought, and assumes that those actions are consistent with long-term maximization of the ability to earn profits. The decision regarding the form in which profits are accepted is a function of the internal organization of the firm and the degree of regulation or public ownership. However, the form in which profits are extracted from the firm generally does not affect the competitive conditions in the industry, the prices charged, the products developed, or the desire to establish a protected position.

The basis for the assumption that all firms maximize their ability to earn long-run profits is the observation that unregulated, regulated, and public firms all take similar actions to establish control of their respective markets. No nineteenth-century monopolist was more solicitous of its market control than the current United States Postal Service. The competitive tactics used by European government telecommunications agencies in the early telephone years were similar to the tactics of Bell and Western Union.

The Creation of Barriers to Entry

A wide variety of expenditures by the firms can be used to create barriers to entry and protect their market position. Previous discussions have emphasized the role of building excess capacity and advertising as barrier creating expenditures.[3] Another important form of barrier-creating investment is research and development to create a patent wall around the basic processes used by the firm. Expenditures for inventions or patents that the firm has no present plans to use but which can be used to prevent the intrusion of other firms fit into this category. A further kind of barrier-creating investment is expenditure on developing legal protection. This includes public relations work for general support of a protected position, specific lobbying and legal expense related to potential legislation and to regulatory hearings, and investment in studies or data that can be used to bolster the firm's position but are unnecessary to the ordinary operations and planning of the firm. Another form of investment in barriers to entry is exclusive dealing contracts. Under the reasonable assumption that most firms would prefer freedom to contract later to a contract binding them to one other firm, an exclusive dealing contract requires the firm to offer better terms than an ordinary contract. The better terms can be considered a barrier to entry in the form of preventing future competitors from access to suppliers, routes, or anything else covered by the exclusive dealing contract.

Barriers to entry can be created by investments of the firm and market share can be obtained by selling at sufficiently low prices. Both are a form of capital that is purchased at the expense of lower current profits and produces higher future profits. The firm faces a complex strategy formulation problem, which requires it to estimate the costs and benefits of investing in market share and barriers to entry capital and to develop long-term plans to reach its optimal position. Although the actual methods for making the plans range from very formal and sophisticated planning models with explicit estimates of the actions of actual and potential competitors to intuitive strategies for developing an industry, all firms that aspire to a dominant position in their industries go through some version of the strategic planning process. Even though firms do not necessarily use formal mathematical models to develop strategy, the strategic planning process can be modeled by the following optimal control problem.

Assume that the industry has a dominant firm which attempts to maximize discounted profits over an infinite time horizon. A competitive fringe exists, either actually in the industry or willing to enter with proper inducement. The competitive fringe enters at a rate proportional to the difference between the limit price and the price set by the dominant firm. The limit price (defined as the price at which no net entry or exit occurs) is equal to the cost of the dominant firm plus a function of the stock of barrier to entry capital. The firm's problem is:

$$\text{Maximize } \int_0^\infty \{[p(t) - c]\,[f(p(t)) - x(t)] - g(b(t))\} e^{-rt} dt$$

$$\textit{subject to: } \dot{x}(t) = k[p(t) - c - B(t)^\alpha]$$

$$\dot{B}(t) = b(t) - \delta B(t)$$

where the following notation is used:

$p(t)$ = the price chosen by the dominant firm

c = the dominant firm's cost per unit, assumed to be a constant

$f(p(t))$ = the industry demand function

$x(t)$ = the output of the competitive fringe

$b(t)$ = the dominant firm's rate of investment in barriers to entry

$g(b(t))$ = the cost function for investment in barriers to entry
r = the discount rate
k = the rate of entry or exit of the competitive fringe
α = the elasticity of the limit price minus cost with respect to the stock of barrier capital
B = the stock of barrier to entry capital
δ = the depreciation rate of B

The following assumptions are made:

A1: $f'(p) < 0$; $2f'(p) + (p - c)f''(p) < 0$
A2: $g(b) > 0$; $g'(b) > 0$; $g''(b) > 0$
A3: $0 < \alpha \leq 0.5$

A1 says that the demand curve is downward sloping and that gross profit at a point in time (ignoring investments in barriers to entry) is a strictly concave function of price. A2 says that the cost of investments in barriers to entry is positive and shows increasing marginal costs. A3 restricts the elasticity of the difference between limit price and cost with respect to the stock of barrier capital to values less than or equal to 0.5. This is done in order to give meaningful determinate results without imposing a complex restriction on $g''(b)$. If $\alpha > 0.5$ and $g''(b)$ is not great enough, the optimal stock of barrier capital may increase without limit.

The control variables are $b(t)$ and $p(t)$. The stock variables are $x(t)$ and $B(t)$. The square brackets give the dominant firm's profit per unit and output, respectively, so that the product of the two is the gross profit at time t. The instantaneous investment in barriers to entry is then subtracted to get the net profit. The firm's current price and the stock of barrier capital determine the change in competitive production at time t. The firm's current investment in barriers to entry and the depreciation rate of existing barrier capital determine the change in barrier capital.

Gaskins's dynamic limit pricing model can be derived as a special case of this one by setting $\delta = 0$ and letting $g(b)$ approach infinity.[4] The stock of barrier capital would then remain at its initial level, because it would not depreciate and the firm would have no incentive to add to it. The limit price would be a constant equal to $c + B(0)^{\alpha}$. The firm's problem would be reduced to finding the optimal price to maximize long-run profits with a given level of barriers to entry. The model could be further reduced to a static limit pricing model as discussed in the early Bain work by letting k approach infinity.[5] In that case, the competitive production would adjust instantaneously to any excess of price over the limit price, and the firm would maximize profits by setting the initial price equal to the limit price and never changing it.

To solve the problem, we may form the Hamiltonian:

$$H = \{[p(t) - c]\,[f(p(t)) - x(t)] - g(b(t))\}\,e^{-rt} + Z_1(t)k[p(t) - c - B(t)^{\alpha}] + Z_2(t)\,[b(t) - \delta B(t)]$$

and get the following necessary conditions:

$$\dot{x}(t) = \partial H/\partial Z_1(t) = k[p(t) - c - B(t)^{\alpha}]$$
$$\dot{B}(t) = \partial H/\partial Z_2(t) = b(t) - \delta B(t)$$
$$\dot{Z}_1(t) = -\partial H/\partial x(t) = (p(t) - c)e^{-rt}$$
$$\dot{Z}_2(t) = -\partial H/\partial B(t) = Z_1(t)k\alpha B(t)^{\alpha-1} + \delta Z_2(t)$$

max H with respect to $p(t)$ and $b(t)$

$$\lim_{t\to\infty} Z_1(t) = 0 \qquad \lim_{t\to\infty} Z_2(t) = 0$$

By A1, $2f'(p(t)) + (p(t) - c)f''(p(t)) < 0$. Consequently the Hessian of H with respect to $p(t)$ and $b(t)$ is negative definite and we can maximize H with respect to $p(t)$ and $b(t)$ by setting $\partial H/\partial p(t) = 0$ and $\partial H/\partial b(t) = 0$. If we also make the following substitutions:

$$\lambda(t) = Z_1(t)e^{rt}$$
$$\mu(t) = Z_2(t)e^{rt}$$

we can rewrite the necessary conditions as the following set of equations:

$$(1)\ \dot{x}(t) = k(p(t) - c - B(t)^{\alpha})$$
$$(2)\ \dot{B}(t) = b(t) - \delta B(t)$$
$$(3)\ \dot{\lambda}(t) = (p(t) - c) + r\lambda(t)$$
$$(4)\ \dot{\mu}(t) = \lambda(t)k\alpha B(t)^{\alpha-1} + \mu(t)(\delta + r)$$
$$(5)\ f(p(t)) - x(t) + (p(t) - c)f'(p(t)) + k\lambda(t) = 0$$
$$(6)\ -g'(b(t)) + \mu(t) = 0$$

By solving equations (5) and (6) for $\lambda(t)$ and $\mu(t)$, respectively, differentiating with respect to time, substituting the result into equations (3) and (4), respectively, we get the following expressions:

$$(7)\ \dot{b}(t) = \frac{\lambda(t)k\alpha B(t)^{\alpha-1} + (\delta + r)\,\mu(t)}{g''(b(t))}$$

$$= \frac{-\alpha B(t)^{\alpha-1}\{f(p(t)) - x(t) + f'\,(p(t))(p(t)-c)\}+g'(b(t))(\delta+r)}{g''(b(t))}$$

$$(8)\ \dot{p}(t) = \frac{-k(r\lambda(t) + B(t)^{\alpha})}{2f'\,(p(t)) + (p(t) - c)f''(p(t))}$$

$$= \frac{r\{f(p(t)) - x(t) + (p(t) - c)f'\,(p(t))\} - kB(t)^{\alpha}}{2f'\,(p(t)) + (p(t) - c)f''(p(t))}$$

Because $\lambda(t)$ is the undiscounted value to the firm of one additional unit of competitive production ($\lambda(t) < 0$), $\mu(t)$ is the undiscounted value to the firm of one additional unit of barrier capital, and $k\alpha B(t)^{\alpha-1}$ is the negative of the partial derivative of $\dot{x}(t)$ with respect to $B(t)$, equation (7) can be interpreted as saying that spending on barriers to entry is increasing whenever the value of the flow of services from barrier capital $((\delta + r)\mu(t)$ is greater than the cost of additional entry times the reduction in the rate of entry from an additional unit of barrier capital. In other words, $\dot{b}(t)$ is positive whenever the net benefit to the firm of a marginal unit of barrier capital is positive. The rate of adjustment of barrier capital to desired barrier capital is inversely proportional to $g''(b(t))$. If $g''(b(t)) = 0$, the adjustment is instantaneous and equation (7) breaks down.

In equation (8), the denominator is negative and k is positive, so $\dot{p}(t) > 0$ whenever $r\lambda(t) + B(t)^{\alpha} > 0$. Since $B(t)^{\alpha}$ is the difference between limit price and dominant firm cost, we may say that price is increasing when the nonentry inducing profit per unit of production is greater than the interest rate times the cost to the firm of one more unit of competitive production. In the special case of $\dot{\lambda}(t) = 0$, we can use equation (3) to say $\dot{p}(t)$ is a positive function of $\bar{p}(t) - p(t)$, where $\bar{p}(t)$ is the limit price (equal to $c + B(t)^{\alpha}$).

In the steady state, $\dot{x}(t) = 0$, $\dot{B}(t) = 0$, $\dot{\lambda}(t) = 0$, and $\dot{\mu}(t) = 0$. By substituting those values into equations (1) through (4), we can treat equations (1) through (6) as a system of six equations with six endogenous variables ($\hat{p}$, $\hat{B}$, $\hat{b}$, $\hat{x}$, $\hat{\lambda}$, $\hat{\mu}$), which characterize the steady-state solution, where the ^ over each variable designates the steady-state value of the variable.

We can solve equation (3) for $\hat{\lambda}$ and equation (6) for $\hat{\mu}$ and substitute the resulting expressions into equation (4). We can then solve equation (1) for $\hat{p}$ and equation (2) for $\hat{b}$ and substitute the resulting expressions into equation (4) to get the following:

$$(9)\ -k\alpha\hat{B}^{2\alpha-1} + r(\delta + r)g'(\delta\hat{B}) = 0$$

Additional relationships which can be developed from equations (1) through (6) include:

$$(10)\ \hat{p} = c + \hat{B}^{\alpha}$$
$$(11)\ \hat{b} = \delta\hat{B}$$
$$(12)\ \hat{x} = f(\hat{p}) + (\hat{p} - c)(f'(\hat{p}) - k/r)$$

Equation (9) gives an implicit expression for $\hat{B}$ as a function only of exogenous variables from which the comparative statics results for $\hat{B}$ can be computed. Equations (10) and (11) express $\hat{p}$ and $\hat{b}$, respectively, as functions of $\hat{B}$, while equation (12) expresses $\hat{x}$ as a function of $\hat{p}$. Since $d\hat{p}/d\hat{B} > 0$ and $d\hat{b}/d\hat{B} > 0$, the variation in $\hat{p}$ and $\hat{b}$ with respect to changes in an exogenous variable will be of the same sign as the variation in $\hat{B}$. From equation (12), $d\hat{x}/d\hat{p} = 2f'(\hat{p}) + (\hat{p} - c)f''(\hat{p}) - k/r < 0$. Consequently, the variation in $\hat{x}$ will be of opposite sign from the variation in $\hat{p}$ and $\hat{B}$. The partial derivative of equation (9) with respect to $\hat{B}$ is $-k\alpha(2\alpha - 1)B^{2\alpha-2} + \delta r(\delta + r)g''(\delta\hat{B})$, which is positive because $\alpha \leq 0.5$ and $g''(b) > 0$. The signs of the partial derivatives of B with respect to each of the exogenous variables are therefore equal to the negative of the signs of the partial derivatives of equation (9) with respect to each exogenous variable. The comparative statics results for the stationary state are summarized in Table 1.

The first column shows that $\hat{x}$ decreases with increases in the rate of competitive response to prices above the limit price and increases with increases in the discount rate. The steady-state value $\hat{x}$ also decreases with α (the effectiveness of barrier capital in preventing entry) and increases with δ (the depreciation rate of barrier capital). These two results are consistent with the intuitive notion that the more expensive it is to create and maintain effective barriers to entry, the higher the competitive production will be. The final entry in the first column shows that $\hat{x}$ will decrease with increases in the cost level.

The second column shows that the steady-state stock of barrier capital increases with the rate of competitive adjustment and with increasing significance of the stock of barrier capital in raising limit price. The

TABLE 1. Comparative statics summary

Exogenous variables	Endogenous variables[a]			
	$\hat{x}$	$\hat{B}$	$\hat{p}$	$\hat{b}$
k	−	+	+	+
r	+	−	−	−
α	−	+	+	+
δ	+	−	−	−
c	−	0	+	0

Source: Calculations in text.

a. Each entry is the sign of the partial derivative of the endogenous variable in the column with respect to the exogenous variable in the row.

stock of barrier capital decreases with increases in the discount rate and the depreciation rate of barrier capital, while it is unchanged with respect to changes in the production cost of the firm. These are intuitively reasonable results. As the rate of competitive response to price above the limit price increases, the firm's incentive is to raise the limit price by investment in barriers to entry, Conversely, as k approaches zero, the firm's incentive to create barriers to entry disappears. An increase in α is equivalent to an increase in the effectiveness of barrier capital in forestalling entry and thus amounts to a reduction in the cost of avoiding entry through creating barriers. Because barriers are treated as a form of capital, the cost of maintaining a given stock of barriers is positively related to both r and δ, accounting for the negative effect of increases in r and δ on $\hat{B}$.

From a policy point of view, the last column of Table 1 is significant because it indicates ways in which the firm may be induced to reduce its expenditure on barriers to entry. Because $\hat{b}$ rises with k, a reduction in the speed at which new firms enter when the price is above the limit price will reduce the expenditures on barriers to entry. Because the limit price includes the effects of the stock of barrier capital, the firm has little incentive to raise the limit price through barrier investments if firms do not respond to prices above the limit price very fast. As k approaches zero, incentive to invest in barriers to entry disappears because no entry takes place regardless of the level of price and barriers to entry. As k approaches infinity, firms enter immediately in response to profitable opportunites, and the dominant firm must invest in barriers to entry if price is to be held above cost. An increase in the response speed of new entrants reduces the value of market share as a form of capital and increases the value of barriers to entry as a form of capital. Consequently we have the result that less wasteful expenditure on barriers to entry would occur if new firms were slower in responding to profitable opportunities.

Because $\hat{b}$ falls with increases in r, an increase in the firm's internal discount rate (a reduction in the firm's planning horizon) will reduce expenditure on barriers to entry. The firm that concentrates on the short run will not find it worthwhile to invest in barriers to entry. Insofar as social policy could affect either k or r, it is doubtful that net social benefits could be obtained from reducing k or increasing r. Although both moves would reduce expenditure on barriers to entry, both would have other less desirable effects, such as reducing flexibility in the economy and restricting useful investments which would only be done by firms with a long time horizon.

Because $\hat{b}$ increases with increases in α, the firm increases its barrier to entry expenditures with increases in the effectiveness of barrier capital. An increase in the effectiveness of barrier capital is equivalent to a reduction in the cost of protection derived from investments in barriers to entry. Similarly, because $\hat{b}$ decreases with increases in δ, the firm decreases its expenditure on barriers to entry with increases in the depreciation rate of barrier capital. An increase in the depreciation rate of barrier capital is also equivalent to an increase in the cost of protection derived from investments in barriers to entry. Consequently any policy that increases the expense of deriving protection from created barriers to entry will reduce the expenditure on barriers to entry.

Although the actual situation is more complex than that described in the model, the results of the model help to clarify the conduct and structure interactions. If there is no technological change or other exogenous changes, we expect to see industries change in structure relatively rapidly in the early years and then become relatively stable as the firms adjust to the steady-state path. Consequently, for industries that have not encountered significant exogenous change in the parameters for many years, the assumption of constant structure which determines the conduct of the industry is quite accurate. However, for industries that are either new or have undergone significant exogenous changes in recent years, the model suggests that the structure should be changing significantly as the industry participants attempt to reach the new optimum industry structure.

One class of exogenous change is change in the political climate. This can take the form of specific changes, such as new laws or new court decisions which affect the ability of the firm to possess market power, or it can be more amorphous changes, such as a perception that the public is less (or more) likely to tolerate market power and may take actions adverse to the firm. A change in political climate adverse to the firm can be interpreted as an increase in the marginal cost of building barriers to entry ($g'(b(t))$). The firm must now spend more money to achieve the same effect as before. From the comparative statics calculations, it can be seen that an increase in $g'(b(t))$ will reduce the steady-state stock of barrier to entry capital, the steady-state price, and the steady-state rate of building barriers to entry, and increase the steady-state competitive market share. Thus it is not necessary to actually destroy barriers to entry through political action in order to reduce them; it is sufficient to increase the cost of building barriers to entry.

A variety of policies can increase the expense of deriving protection by decreasing the effectiveness of barrier capital or by increasing the rate of depreciation of barrier capital. The Sherman Act's prohibition of monopolizing, which has been interpreted to prohibit a variety of actions that contribute to monopoly power, provides an example of existing policy that reduces barrier to entry expenditures. Clayton Act restrictions on requirements contracts and exclusive dealing provide another example. Antitrust limitations on the use of patents to restrict entry also reduce the incentive to invest in that form of entry protection. Reducing barriers investment reduces wasteful expenditures and also leads to more competition through its effect on the stock of barrier capital and the amount of new entry.

The previous discussion suggests that conduct remedies in antitrust cases may be more effective than has generally been conceded by industrial organization economists. Many economists have lamented the reluctance of the courts to impose structural remedies in antitrust actions. Because they perceive the source of competitive problems in undesirable industry structures, conduct remedies are seen as treating symptoms rather than the disease and likely to be ineffective. However, if the industry will change its own structure in response to changing external circumstances, a conduct remedy can be effective. Conduct restrictions that eliminate some of the established barrier capital and increase the cost of building additional barriers may be more efficient in restoring competitive structure than a dissolution of the offending firm or firms.

The conduct remedy is slower and less dramatic than the structural remedy, but it removes the incentive to recreate the offending structure and reduces the transitional losses in efficiency during the reorganization period. The choice of remedy obviously varies with the circumstances of each case, but conduct remedies should not be dismissed out of hand as inferior to structural remedies.

An additional implication of the model is that trends in concentration and barriers to entry should be taken into account in assessing profit figures. An industry showing current high profits, but with declining market share of the dominant firm and declining barriers to entry, is earning a lower level of profits than the reported level because part of its profits are coming from selling its market power. Similarly a firm with increasing market share and barriers to entry is earning higher profits than the reported profits would indicate because part of the profits are going to purchase additional market position. An accurate assessment of the relationship between market structure and profitability should therefore include consideration of changes in market structure variables as well as in their absolute level.

New Entry Via Exogenous Technological Progress

Extensive theoretical and empirical studies of the relationship between industry structure and technological progress have been undertaken in recent years.[6] Most of these studies have been concerned with evaluating the effect of different industry structures on the rate of technical progress. The theoretical studies have explored the varying incentives to undertake research and development under monopoly, competition, and intermediate degrees of rivalry. The empirical studies have used case studies and statistical methods to define the relationship between industry structure and various measures of technical progress, such as patents, significant inventions, or research and development spending. The results of the studies have not been entirely definitive but suggest that the relationship between industry structure and technological progress is complex. No single industry structure has emerged as clearly optimal for stimulating research and development. The studies suggest that some form of oligopoly is superior to either pure competition or pure monopoly for encouraging technical progress, but that forces other than industry structure are more significant than industry structure in determining the rate of technical progress for any particular industry.

Most of the work on the relationship between technological progress and industry structure has assumed that the action takes place within a well-defined industry with a given industry structure. Progress is generally viewed as a reduction in the cost of producing an existing product. Much less work has been done on how technological progress changes industry boundaries and the structure of an industry. Yet much of the significance of technological competition is lost when the focus is restricted to changes that do not affect either the industry boundaries or the industry structure. In telecommunications and in many other industries, technological progress has been an important cause of changes in barriers to entry and market shares.

The idea that technological progress is a significant form of interindustry competition is hardly new. In an often quoted passage Joseph Schumpeter stated:

> But in capitalist reality as distinguished from its textbook picture, it is not that kind of competition which counts but the competition from the new commodity, the new technology, the new source of supply, the new type of organization (the largest-scale unit of control for instance)—competition which commands a decisive cost or quality advantage and which strikes not at the margins of the profits and the outputs of the existing firms but at their foundations and their very lives. This kind of competition is as much more effective than the other as a bombardment is in comparison with forcing a door, and so much more important that it becomes a matter of comparative indifference whether competition in the ordinary sense functions more or less promptly; the powerful lever that in the long run expands output and brings down prices is in any case made of other stuff.[7]

Schumpeter's view of innovation as a determinant of industry structures has received little attention relative to his hypothesis that monopoly power is a positive influence on technical progress, which has been explored extensively. The assumption of constant industry structure in much industrial organization work has caused researchers to focus on the effect of industry structure on technological progress rather than on the effect of technological progress on industry structure.

One important study that did explicitly examine the impact of technological progress on market structure is Almarin Phillips's study of the aircraft industry.[8] Phillips studied the effect of technological development on the structure of the aircraft industry and concluded that progress had a concentrating effect. Research and development was oriented toward applying the basic technology developed for military aircraft to commercial aircraft. Even though the fundamental advances had already been made for military aircraft, the application to commercial aircraft entailed considerable risk and opportunity for management error. Because the largest firms had more capability to absorb the losses resulting from any particular error without endangering the survival of the firm, the progress had a concentrating effect. In Phillips's view, the rate of progress in a particular industry is largely determined by the progress in the exogenous science related to that industry, and the higher the rate of progress, the more tendency toward concentration arises. Phillips's conceptual framework was important for emphasizing the significance of exogenous scientific advance on the level of research performed by a particular industry and for emphasizing the impact of technological change on market structures.

This section and the following one build on the ideas developed by Schumpeter and Phillips and broaden them to include an array of interindustry competitive effects that arise from technological progress. This section examines changes in industry structure that result from an outside technology becoming economically competitive with existing technology in the industry. The issues may be focused by reference to the well-known story of the displacement of mechanical calculators by electronic calculators. Before 1960 a well-defined mechanical calculator industry existed. Electronic calculators were technologically feasible but so expensive relative to mechanical calculators that they were not competitive with mechanical calculators. During the 1960s rapid technological progress in electronics technology brought continual reductions in the cost of electronic calculators, while the cost of mechanical calculators remained relatively constant. This progress induced established electronics companies such as Texas Instruments and Hewlett-Packard to develop products competitive with mechanical calculators. Continued reduction in the cost of electronic versus mechanical calculators eliminated mechanical technology as a viable option and caused the original mechanical manufacturers to switch to electronics technology in order to remain in the market for calculators. The displacement of mechanical by electronics technology overcame the previous barriers to entry in the market and established the new entrants with sophisticated electronics capability as the market leaders rather than the original leading firms.

The calculator example is a case of interindustry competition. Rapid technological change in one industry (electronic components) allowed its participants to enter a new industry (calculators) without the ordinary cost disadvantages because the superior technology was better known by the entrants than by the established firms. The entrants still faced some entry costs because of the need to design a product and establish marketing channels, but they did not have to learn the technology of the established firms. The case of technological progress in one industry leading to a change in technology in another industry is quite common. In some cases the change in technology causes significant changes in industry structure. In other cases the new technology is adopted by the existing firms without new entry and without upsetting the established industry structure. Which pattern occurs is important for assessing the significance of technological change as a form of interindustry competition. The following simplified model helps to clarify the impact of exogenous technological progress on industry structure.

Assume that a particular industry is controlled by a pure monopolist with constant returns to scale and an average cost of production equal to c_1 per unit. The industry has a straight line demand curve given by $p = a - bq$. The monopolist has a patent or other means of protection on its technology so that it can price without regard to potential entry. It then follows from standard profit maximizing calculations that the initial price and quantity are given by $p_0 = (a + c_1)/2$ and $q_0 = (a - c_1)/2b$. The profit margin (profit per unit) is given by $M_0 = (a - c_1)/2$ and the total profit is given by $\pi_0 = (a - c_1)^2/4b = M_0^2/b$.

A second industry exists which uses a different technology. This technology is undergoing continuous improvement through research and development directed to products of the second industry. However, the second technology can also be used to produce the product of the first industry at a cost given by $c_2e^{-\alpha t}$, where c_2 is enough greater than c_1 that the second technology is not competitive at time zero and $\alpha > 0$. Because the barriers to entry in industry 1 are technology specific, firms from industry 2 may enter industry 1 if they use technology 2. If the monopolist of industry 1 wants to convert to technology 2, it must spend R per year for research and development in order to attain the same cost level as firms in industry 2.

Case 1: Industry 2 Is Competitive

In this case there are many potential entrants with access to technology 2. Industry 2 first becomes a factor in industry 1 when the costs of production via technology 2 decline to the level of the monopoly price, that is when $c_2e^{-\alpha t} = p_0 = (a + c_1)/2$. Because of the large number of potential entrants, entry will occur until the price is brought down to the level of technology 2 costs. Thus after the time that $c_2e^{-\alpha t} = p_0$, the price will be given by $p(t) = c_2e^{-\alpha t}$. The monopolist's choice is to sell q_0 units at the new price or a larger number of units at that same price. The monopolist will thus find it profitable to expand and prevent entry so long as $c_2e^{-\alpha t} > c_1$. If $R > 0$, the original monopolist is at a cost disadvantage relative to the new entrants after $c_2e^{-\alpha t} < c_1$ and chooses to go out of business. We thus see the monopolist maintain 100 percent of the market while the price and profits gradually decline and then go out of business. The monopolist's decline is not caused by lack of foresight but simply by the changing cost conditions over which it has no control.

If $R = 0$, the original monopolist can convert to technology 2 without research and development cost. It then has the same costs as the new entrants. The industry will show competitive performance, but the market shares are indeterminate. The original firm may continue to sell all of the output at the competitive price and make zero profits, or it may exit. Because of the large number of potential entrants with access to technology 2, the social performance of the industry is not dependent on the ability of firm 1 to convert to the new technology. Even if the original monopolist remains the dominant firm, its market power has declined to zero because of the large number of potential entrants with technology 2.

In the more realistic case in which entry occurs over a period of time, the original monopolist will make the greatest profit by gradually selling its market share. It is not limited to selling at a price determined by the costs of technology 2 because entry requires time. The monopolist will find it profitable to maintain its price above the costs of technology 2 and gradually exit from the market as new firms come in. Consequently it will go out of business and will not reenter even with $R = 0$.

The monopolist with foresight will generally find it profitable to build nontechnological barriers to entry prior to the actual entry of firms from industry 2. Regulation is a particularly attractive option in this situation, because the monopolist can promise to maintain current prices (or even reduce the current prices) in exchange for entry protection. The monopolist can easily demonstrate that the initial new entrants have higher costs than the monopolist and thus that inefficient production will result from allowing entry. Because most regulatory agencies are generally concerned with the profit level only at the time a price increase is requested, the regulators will be unlikely to examine the profit level carefully if no increases are requested. The regulatory solution will appear to be socially beneficial to observers with no knowledge of the feasibility of the alternative technology. The industry will show a price decrease immediately after the regulation is imposed and then steady prices below the monopoly level. "Wasteful and uneconomic entry" of the second firm with initially higher cost technology will be avoided.

If regulation is obtained, the original monopolist will remain the industry monopolist. If it has low-cost access to technology 2, it will switch to that technology soon after the time that technology would have become dominant in the absence of regulation. If access to technology 2 is expensive or impossible to obtain, the monopolist will continue using the original technology under regulation even when technology 2 is the low-cost technology. If factors not included in the model, such as learning by doing or technology-specific capital, are important, the regulated monopolist will switch to technology 2 later than under competition even if it has free access to the new technology.

Case 2: Industry 2 Is a Monopoly

The industry 2 monopolist will enter industry 1 as soon as the costs of technology 2 are below the price set by the industry 1 monopolist, as in the previous case. However, in this case, firm 2 will see the residual demand curve (demand curve minus the output of the industry 1 monopolist) as its demand curve and maximize profits by setting its marginal cost equal to the marginal revenue derived from that residual demand curve. With a straight line demand curve, the entrant will produce half the number of units necessary to bring the price down to its level of costs. Thus the relevant concern for the monopolist is the relative cost of adding one unit to its own production or having another firm add one-half unit to the industry production. Allowing entry will mean a higher price and lower quantity for the monopolist than deterring entry. At the monopoly level price it is profitable to deter entry. At the competitive price, deterrence is unprofitable because the price reduction for deterrence is not compensated for by the profit margin on the additional units of production.

Because deterrence is profitable at the monopoly level and unprofitable at the competitive level, we can find the optimal deterrence level of output for the original firm, q_1, by setting the marginal cost of adding a small increment to output to deter entry equal to the marginal cost of the entry that would result without the deterrence. The cost of adding one unit to deter entry when output is q_1 is given by $c_1 - (a - 2bq_1)$. Without deterrence, entry would be one-half of the needed deterrence addition, resulting in a price reduction of $\frac{1}{2}b$ for all q_1 units produced by the monopolist. The cost of initial entry is thus $\frac{1}{2}bq_1$. Setting the two expressions equal gives $q_1 = \frac{4}{3}q_0$. Thus it is profitable for the firm to deter entry until its output is equal to $\frac{4}{3}$ of the original monopoly output and allow entry after that time. At the maximum deterrence point, the profit margin will have been reduced to $\frac{2}{3}$ of the original profit margin and profits will be at $\frac{8}{9}$ of the original level.

When $c_2e^{-\alpha t} < c_1 + 0.66M_0$, it is profitable to allow entry even though the entrant has higher costs than the established firm. So long as the dominant firm expects the entrant to maximize its potential profits from the residual demand curve, it will find it profitable to gradually reduce its output as the new firm enters. For any level q_1 of original firm output, the entrant will maximize profits by choosing a price and quantity given by:

$$p = ((a - bq_1) + c_2e^{-\alpha t})/2 \qquad q_2 = ((a - bq_1) - c_2e^{-\alpha t})/2b$$

Knowing how the entrant will respond to its output, the original firm can maximize profit by setting its output at:

$$q_1 = q_0 + (c_2e^{-\alpha t} - c_1)/2b$$

so long as some entry occurs and the costs of technology 2 are above the costs of technology 1. This causes the original firm to gradually decrease its output from $\frac{4}{3}q_0$ at the time of first entry to q_0 at the time the costs of the two technologies reach equality. This contrasts with the competitive case in which the monopolist gradually expands output from q_0 at the time of first potential entry to $2q_0$ (the competitive industry output) at the time the costs of the technologies reach equality. In the monopoly case, the entrant's output steadily increases from zero when entry first occurs to $\frac{1}{2}q_0$ when the costs of the two technologies reach equality. At that point total output has been increased to $\frac{3}{2}$ of the monopoly level and the profit margin has been cut in half. Profits for the original firm are one-half of the original level and profits for the entrant are one-quarter of the original monopoly profits in the industry.

So long as the costs of technology 2 are above those of technology 1, the von Stackelberg leader-follower reactions are quite reasonable. There is no incentive for the entrant to expand or threaten to expand its output beyond that given, because to do so would invite additional expansion by the established firm. Because the established firm has lower costs than the entrant, it is better able to weather a price war and thus is able to maintain its role as industry leader. However, after costs reach equality, the entrant becomes the low-cost firm if R is positive. If R is zero, the original firm can switch to the new technology without cost and we then have a duopoly with equal costs and indeterminate output division.

If $R > 0$, it is likely that the leader-follower roles will be reversed after the cost of technology 2 declines to the cost of technology 1. The entrant becomes the low-cost firm and maximizes profits taking account of the entire demand curve, while the original monopolist is reduced to a follower that operates on the residual demand curve. The switch in roles causes a discontinuity in the output of the two firms. The entrant's output, which had been steadily growing toward $\frac{1}{2}q_0$, jumps to q_0. The original monopolist's output, which had been steadily shrinking toward q_0, drops to $\frac{1}{2}q_0$. Total output remains the same immediately after the switch in roles. After the equality point, we must consider the costs of switching to the new technology. First, assume that the costs are very high so that the old firm must remain with technology 1. Define $A(t)$ as the cost advantage of the entrant over the original firm: $A(t) = c_1 - c_2e^{-\alpha t}$. Assuming the entrant acts as leader and the original firm as follower, we can compute the path of price and quantity as:

$$q_1 = \tfrac{1}{2}q_0 - A(t)/2b \qquad q_2 = q_0 + A(t)/b$$
$$p = (p_0 + c_1 - A(t))/2 = (p_0 + c_2e^{-\alpha t})/2$$

As the cost advantage of the new technology increases over time, the entrant expands its output and the original firm contracts its output. Total industry output increases and the original monopolist loses both volume and unit profit margin as technological progress continues.

$A(t)$ is bounded by c_1. If $c_1 < M_0$, the original firm is never driven out of the market even if the costs of technology 2 decline to zero. If $A(t)$ reaches M_0, the entrant sets q_2 equal to $2q_0$ and price equal to c_1 and the original firm disappears from the industry. At that point, the entrant has 100 percent of the output but is not a pure monopolist because monopoly pricing would bring the original firm back into the industry. The optimal path for the entrant is to hold output constant from the time $A(t) = M_0$ until $A(t) = 2M_0$ with price constant at c_1. If $A(t)$ reaches $2M_0$, the entrant can switch to pure monopoly pricing without regard for potential entry by the original firm.

If the costs of switching technologies are not prohibitive, the original firm may switch rather than be driven out of the industry. If it switches, it achieves the same marginal cost as the entrant but must pay a fixed cost R per year for research and development (or for royalties) in order to gain access to the new technology. If switching occurs, both firms will modify their output to account for the reduced marginal costs of the original firm. The profits for firm 1 without switching are given by:

$$\pi_1 = (a - c_1 - 2A(t))^2/16b = (M_0 - A(t))^2/4b$$

After switching, the quantity, profit margin, and net profits for firm 1 are given by:

$$q_1^s = (2M_0 + A(t))/4b$$
$$M_1^s = (2M_0 + A(t))/4$$
$$\pi_1^s = (2M_0 + A(t))^2/16b - R$$

The net benefit of switching technologies is given by the difference in the two profit expressions:

$$\pi_1^s - \pi_1 = (12M_0A(t) - 3A(t)^2)/16b - R$$

When $A(t) = M_0$ (at which time firm 1 exits without switching) the net benefits of switching can be written:

$$\tfrac{9}{16} M_0^2/b - R = \tfrac{9}{16}\pi_0 - R$$

Thus if R is less than $\frac{9}{16}$ of the original monopoly profits, firm 1 will find it advantageous to switch technologies prior to being driven out of the market. The advantage of switching is equal to $-R$ when $A(t) = 0$ and increases in $A(t)$. The firm will switch when the profit difference expression is first equal to zero. The switch point will cause a discontinuity in the market shares as the original monopolist increases its output and the entrant decreases its output to account for the new equality of marginal costs.

Firm 2's profit margin in the absence of switching is $\frac{1}{2}M_0 + \frac{1}{2}A(t)$. With switching, it is reduced to $\frac{1}{2}M_0 + \frac{1}{4}A(t)$. The ability of firm 1 to switch technologies increases total output and decreases the price. It also raises average industry costs, because the costs of firm 1 are higher than those of firm 2 due to the research cost R.

The previous discussion assumes that the firms act independently without collusion. This case offers ample opportunity for collusion, which can increase total industry profits and probably avoid antitrust prosecution as well. Because the firms are initially in different industries and the competition is over basic technology, the firms could arrange a technology exchange or patent licensing agreement as the basis for a cooperative solution without making any agreement on prices or market shares. The collusive industry profit maximizing solution would be to maintain output at q_0, price at p_0, and all production with technology 1 for the period in which the costs of technology 1 are below the costs of technology 2. When technology 2 becomes cheaper, all output would be switched to technology 2, the price would be $p_0 - \frac{1}{2}A(t)$, and the quantity would be $q_0 + A(t)/2b$.

Several possibilities exist for potentially legal collusive agreements in which both firms earn greater profits than under the noncollusive solution. One method is for firm 1 to purchase the rights and technical assistance to produce its product undr technology 2 from firm 2. The royalty payments could be chosen to gradually increase and remain a little above firm 2's expected profits from entry or they could be a fixed sum equal to the discounted value of firm 2's expected profits. Firm 1 would remain the monopolist and use technology 1 until $A(t) = 0$, then switch to technology 2 and remain a monopolist. It would appear to be a progressive firm adopting the best possible technology over time. Antitrust questions would be unlikely to arise so long as the technology was purchased before firm 2 made any definite efforts to enter the industry. Firm 1 would be simply purchasing rights to advanced technology from a firm in a different industry.

A second possibility is for firm 1 to agree to purchase its needs from firm 2 after the time that $A(t) = 0$. Firm 1 would then be an integrated monopolist prior to the time that $A(t) = 0$ and a retailing monopolist after that time, while firm 2 would be the manufacturing monopolist. In a world of foresight and certainty, the agreement could include specific quantity and price provisions designed to reach the industry profit maximum position. So long as the agreement simply called for firm 2 to supply a specific number of units at a specific price, it would be immune to antitrust attack. If demand were uncertain, the agreement might specify the price but leave the quantity uncertain. In that case, firm 1 would see the price paid to firm 2 (including monopoly profits) as its cost and would contract output compared to the monopoly position. Output would be below the industry profit maximum level and not all potential profits would be achieved.

As in the case of competition in industry 2, the original monopolist in this case is likely to attempt to build nontechnological barriers to entry if they can be obtained at reasonable cost. Regulation remains an attractive option because the monopolist is attempting to maintain an established position against a threat to that position. Other forms of barrier-creating expenditures, including investments in excess capacity, may allow the monopolist to deter entry. With regulation, the monopolist may be able to maintain control while using the old technology regardless of the cost advantages of the new technology. Without legal protection, the original monopolist is unlikely to maintain control unless it can switch to the new technology with little cost disadvantage relative to the potential entrant.

Changes in Industry Boundaries

This section explores the role of technological progress in creating new competition through changes in industry boundaries. It focuses on the demand interactions of related but nonidentical products as their relative prices change. In contrast to the preceding section, no explicit entry into an existing industry takes place. Rather, an existing high-priced specialized product undergoes technological change which moves it from a specialized niche to a mainline competitor of an established product.

Historical examples of this process include the competition of telephones with telegraph and the competition of trucks with railroads. In both cases the new products (telephones and trucks, respectively) were first introduced to meet specialized needs. They lacked the ability to provide significant competition to the established products. The initial pricing of the new services was well above the comparable prices of the established services. This caused the new services to only be used for functions that the old services performed poorly. It also prevented any significant effect on the demand for the old product from changes in the price of the new product. Thus the products were properly seen as operating in different markets. Over time, the new products enjoyed greater technological progress than the old products. As their prices were reduced, they became mainline competitors to the old products and even displaced much of their business. The reduction in price of the new products brought them into competition with the established products without explicit entry.

The demand side competition is a more subtle form of increased competition than direct entry. If the changes are slow and the products are highly differentiated, the established firm may not even recognize the increased competition until it is quite strong. Collusion is more difficult because of the differentiated nature of the products. Customers have definite preferences for one product or the other and it is difficult to reach a market-sharing agreement. Regulation provides less protection in this case than the previous one because no actual entry is taking place. Although an established firm may be able to persuade the regulatory agency to limit the competition from the new product, it does not have the kind of automatic protection that it has against entry. A change in regulation is needed to bring the new product under the control of the regulatory agency.

In some cases regulation may even hinder the established firm in fighting new competition. The basic defense of an established firm consists of entry into the new product, or the threat of entry to gain concessions from the potential competitor. If the product is defined to be outside the existing regulatory boundaries and existing firms are constrained to remain inside those boundaries, the new competitors rather than the old competitors may gain from the existence of regulation. If the regulatory agency requires the continuation of pricing patterns established prior to the competition of the new product, new firms may gain an added advantage. This, for example, appears to have been true in the competition of trucks and railroads.

In the following model two goods are postulated which fulfill the same primary function but have different characteristics. An example would be telegraph service and long-distance telephone service. Both provide rapid communication between distant points. Telegraph service provides a written record of the communication. Telephone service provides instantaneous two-way communication with the personalized aspect of actual voice transmission, but no written record. If telegraph service is cheaper for basic communication than telephone, all users who prefer the telegraph characteristics and some who prefer the telephone characteristics will choose telegraph. Telephone will be restricted to those customers who place a high value on the specific telephone characteristics. If telephone service is much more expensive per message than telegraph, only customers with great preference for the telephone characteristics will choose telephone, and the telegraph company will not see the telephone service as competitive. As the price of telephone service comes down to below the telegraph price, all customers who prefer the telephone characteristics will choose telephone, and only those who prefer the special telegraph characteristics will choose telegraph.

Assume that there are three goods in the economy, x, y, and z. The goods y and z have the same fundamental function but different characteristics. The good x represents a composite of all other possible goods. Half of the population prefers the characteristics of y and half prefers the characteristics of z. The degree of preference of each person is represented by a random variable m which is distributed as the negative exponential distribution beginning at 1; $f(m) = e^{1-m}$. The variable m ranges from 1 to infinity and is the weight to be applied to the preferred good in the utility function. A person who prefers the characteristics of y will be indifferent between one unit of y and m units of z; a person who prefers the characteristics of z will be indifferent between one unit of z and m units of y. Because ${}_1\int^2 e^{1-m}dm = 0.63$, 63 percent of those who prefer y will choose z if the price of y is more than twice the price of z and 63 percent of those who prefer z will choose y if the price of z is more than twice the price of y.

Assume that the composite utility function of people who prefer y can be written as:

$$U(x,y,z) = x^{\alpha}(my + z)^{\beta}$$

and their aggregate income is equal to I. Except in the case of $m = p_y/p_z$ in which the division between y and z is indeterminate, individual consumers will choose only y or z, not both. Consequently, we can simplify the utility function by the substitution $q = my + z$. The price of the good q is given by $p_q = \min(p_y/m, p_z)$. We can then derive the demand function for q by straightforward maximization of the utility function subject to the budget constraint $p_x x + p_q q = I$. The resulting demand function is:

$$q = \beta I/(\alpha + \beta)\, p_q = \gamma I/p_q \qquad \text{where } \gamma = \beta/(\alpha + \beta)$$

To develop the demand functions for y and z, we can decompose q according to the size of m and the prices p_z and p_y. If $p_y < p_z$, all demand from those who prefer y goes to y and none to z. Assume $p_y = \delta p_z$ with $\delta >$ 1. Then those customers with $m < \delta$ will purchase z instead of y. Those with $m > \delta$ will purchase y. The respective demand functions can consequently be found by integrating over the appropriate ranges of the density function of m. The demand functions are:

$$\begin{aligned} z &= {}_1\!\int^{\delta} e^{1-m} \gamma I/p_z dm = (\gamma I/p_z)[1 - e^{(1-p_y/p_z)}] \\ &\phantom{= {}_1\!\int^{\delta} e^{1-m} \gamma I/p_z dm} = (\gamma I/p_z)(1 - \omega) \\ &\text{where } \omega = e^{(1-p_y/p_z)} \\ y &= {}_\delta\!\int^{\infty} e^{1-m} \gamma I/p_y dm = \gamma I\omega/p_y \end{aligned}$$

The same procedure can be applied to those who prefer the characteristics of z with the substitution $q = y + mz$ and a resulting price $p_q =$ $\min(p_y, p_z/m)$. For $p_y = \delta p_z$ as before but with $\delta < 1$, we get the demand functions for those who prefer z:

$$\begin{aligned} y &= {}_1\!\int^{1/\delta} e^{1-m} \gamma I/p_y dm = (\gamma I/p_y)(1 - \mu) \\ z &= {}_{1/\delta}\!\int^{\infty} e^{1-m} \gamma I/p_z dm = \gamma I\mu/p_z \qquad \text{where } \mu = e^{(1-p_z/p_y)} \end{aligned}$$

The total demand functions for y and z are the sum of the demand functions for those who prefer y and the demand functions for those who prefer z. For $p_y > p_z$ the total demand functions are:

$$\begin{aligned} y &= \gamma I\omega/p_y \\ z &= \gamma I/p_z + (\gamma I/p_z)(1 - \omega) \\ &= (\gamma I/p_z)(2 - \omega) \end{aligned}$$

For $p_y < p_z$ the demand functions are:

$$\begin{aligned} y &= (\gamma I/p_y)(2 - \mu) \\ z &= \gamma I\mu/p_z \end{aligned}$$

When $p_y > p_z$, the own price elasticity of z and the cross price elasticity of z with respect to p_y are given by:

$$(\partial z/\partial p_z)(p_z/z) = -[1 + p_y\omega/p_z(2 - \omega)]$$
$$(\partial z/\partial p_y)(p_y/z) = p_y\omega/p_z(2 - \omega)$$

The own price elasticity of y and the cross price elasticity of y with respect to p_z are given by:

$$(\partial y/\partial p_y)(p_y/y) = -(1 + p_y/p_z)$$
$$(\partial y/\partial p_z)(p_z/y) = p_y/p_z$$

For $p_y < p_z$, the elasticity formulas are the same with the role of y and z reversed.

Assume that both y and z are produced with a constant returns to scale technology. Initially the cost of y is much higher than the cost of z, but y undergoes steady cost reducing technological progress while the cost of z remains constant. To allow numerical examples, assume that the initial cost of producing y is five times the cost of producing z and that the cost of y declines 5 percent per year: $c_y(t) = 5c_z e^{-.05t}$.

Case 1: Perfect Competition in Both y *and* z

At time zero, $p_y = 5p_z$ and the quantity demanded of the two goods is given by:

$$y = .0183\gamma I/p_y = .0037\gamma I/c_z$$
$$z = 1.9817\gamma I/p_z = 1.9817\gamma I/c_z$$

Sales of y in physical terms are 0.2 percent of the combined $y + z$ market. In value terms y has 0.9 percent of the combined market. The own price elasticity of z is equal to -1.046 and the cross price elasticity of z with respect to p_y is 0.046. The own price elasticity of y is -6 and the cross elasticity of y with respect to p_z is 5.

The low cross elasticity of z with respect to p_y at time zero means that producers of z are unlikely to recognize y as specifically competitive. The cross elasticity is less than 5 percent of the own elasticity and y is simply one of the many goods whose price has a minor effect on the sales of z. From y's perspective, the cross elasticity with the price of z is a large proportion of y's own elasticity. Thus y sees a very elastic demand and specific dependence on the price of z. The makers of y see their product as a superior alternative to z for which customers are willing to pay a substantial price premium.

As technological progress brings down the cost of producing y, y gradually increases its market size and z gradually contracts its market size. The cross elasticity of z with respect to the price of y increases as y becomes competitive for a wider range of uses. When the price of y declines to twice the level of the price of z, y's share of the combined market is 10.1 percent in physical terms and 18.4 percent in value terms. The cross elasticity of z with respect to p_y has increased to 0.451. When the prices reach equality, each good has 50 percent of the combined market and the cross elasticity of z with respect to p_y is 1.0. When the price of y declines to half the price of z, the share of z in the combined market drops to 18.4 percent of total value. Thus y is the dominant market and z is a declining industry. As y increases its dominance of the market, its cross elasticity with respect to the price of z diminishes and it sees the two products as becoming less competitive.

The pattern of industry change is shown in Figures 1 and 2. Figure 1 shows the demand for both y and z over time. Figure 2 shows the pattern of cross elasticities between the two products. The different characteristics of the two products prevent them from becoming highly substitutable at any time. Both products exist in the market even with large price differentials because of the differences in customer preferences. Especially when the price differential is large, small changes in price have little effect on the relative market shares. Yet technological progress over a long period of time allows one product to largely displace the other.

Case 2: Monopoly with Imperfect Barriers to Entry

Assume that separate monopolists control products z and y. Each monopolist has high but not perfect barriers to entry. Specifically assume that the limit price for each good is equal to twice the cost of that good. All other assumptions of Case 1 continue in effect. Three options exist for the firms: independent pricing, collusion or merger, and predatory pricing.

With independent pricing, each firm computes its maximum profit position from its cost and elasticity of demand without considering its interaction with the other firm. As in Case 1, assume that $c_y(t) = 5c_z e^{-.05t}$. The monopolist of z cannot charge above $2c_z$ without attracting entry. So long as the own price elasticity of demand for z is less in absolute value than 2, the monopolist will practice limit pricing with $p_z = 2c_z$. The elasticity of demand for z is less than 2 whenever $p_z < p_y$. Conse-

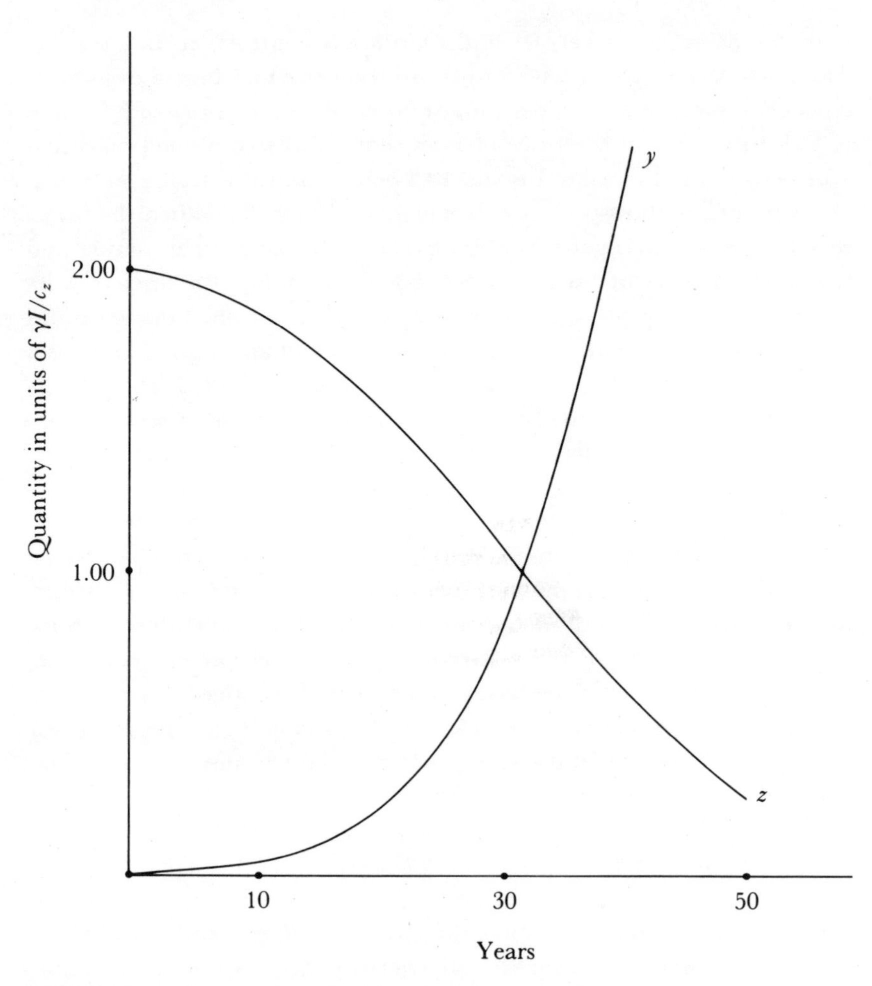

FIGURE 1 Quantity demanded with competition

quently the initial restraint on z's pricing freedom comes from potential entrants, not from cross elasticity with product y.

With $p_z = 2c_z$, the initial own price elasticity of y is -3.5 at the competitive price ($p_y = 5c_z$) and increases in absolute value with increases in p_y. Thus limit pricing by y would be above the maximum profit level. The maximum profit position for y can be found by setting y's marginal revenue equal to its marginal cost and solving for p_y in terms of c_z. This procedure gives $p_y = 6.53c_z = 1.31c_y$ compared with a limit pricing level

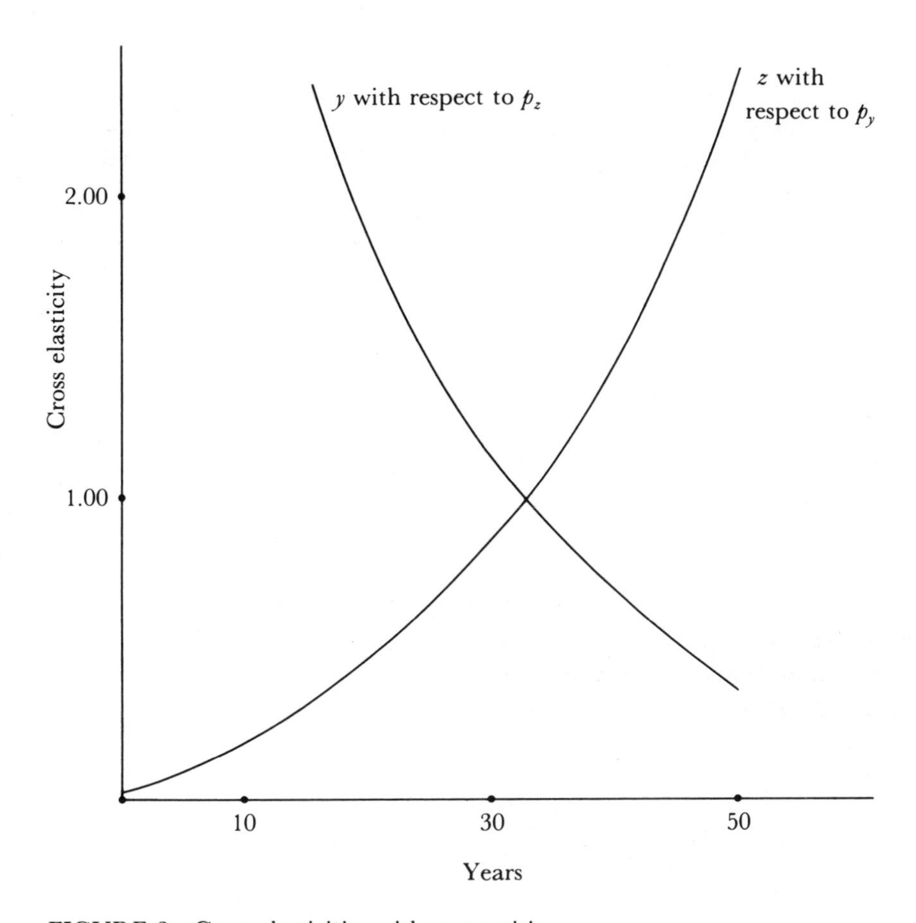

FIGURE 2 Cross elasticities with competition

of $p_y = 10c_z$. The initial quantity and profit for the respective monopolists is given by:

$$y = .0158\gamma I/c_z \qquad \pi_y = .0242\gamma I$$
$$z = .9483\gamma I/c_z \qquad \pi_z = .9483\gamma I$$

As progress occurs in y, its own elasticity of demand gradually decreases (in absolute value) toward 2, while the own price elasticity of z gradually increases toward 2. The monopolist of z will not see any incentive to cut its price below the limit price until the cost of y and z reach equality. Prior to that time, the monopolist is more tightly constrained by potential entrants than by cross elasticity with product y. The monopolist for y will gradually increase its profit margin as its elasticity decreases and will reach the limit price at the point of cost equality. The market share and profits of y will gradually increase, while those of z gradually decrease. Total industry profits decline slightly as y increases its production and then increase again as y increases its profit margin to the level of the limit price. After the cost of y declines to the level of the cost of z, the roles are reversed and y is constrained by potential entrants, while z is constrained by the elasticity of demand. The path of profits for the respective monopolists and for the combined industry is shown in Figure 3.

Greater total profits can be made by collusion or merger. Even though the initial cross elasticity of z with respect to the price of y is very low, the production which y obtains is taken away from z. The low profit margin chosen by y because of its high perceived own elasticity of demand reduces total industry profits below the maximum level. Increasing the price of y to the limit price reduces y's profits but increases total industry

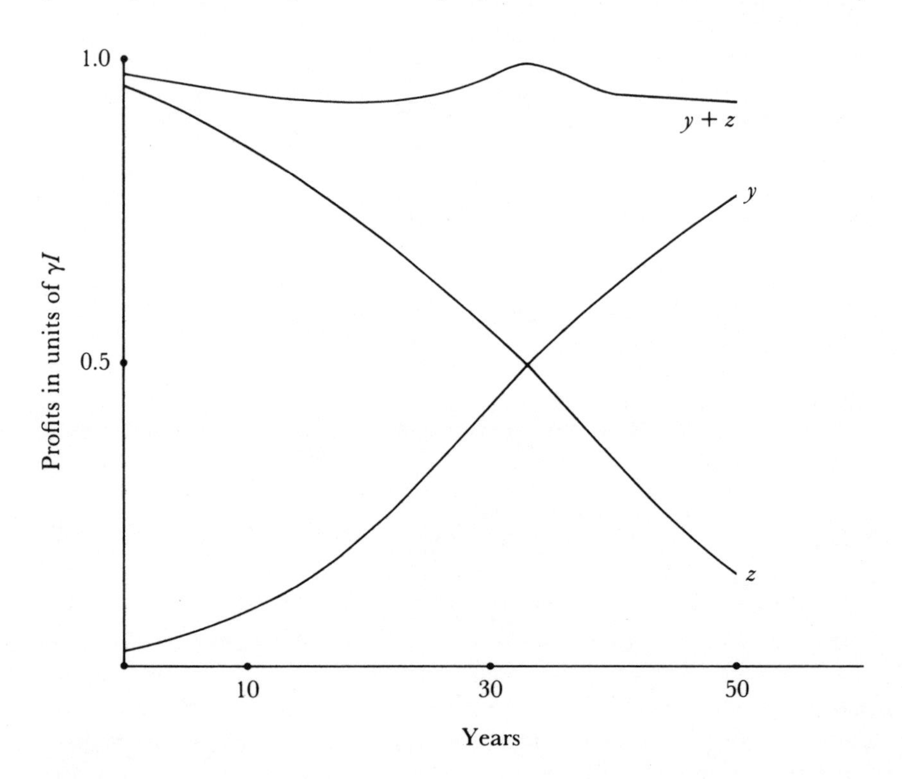

FIGURE 3 Profits with monopoly

profits. With a limit price equal to twice the level of costs, maximum industry profits can be made by pricing both *y* and *z* at twice their respective costs.

Collusion may be difficult to achieve because side payments are required. Under the maximum industry profit scheme, *y*'s profits are lower at all times before cost equality than they would be under independent pricing. Thus *z* must provide some compensation to *y* to induce it to maintain the high price. The easiest way to accomplish the collusion is through a merger. Antitrust problems could be avoided if the merger is undertaken very early, while *y* is a tiny firm with only marginal competitive impact on *z*. If both firms have perfect foresight, a merger can be expected because it is in the interests of both to avoid direct competition and collusion with side payments is difficult to achieve with or without antitrust constraints. However, in the case of some uncertainty regarding the future, merger terms may be difficult to agree upon. If *y* correctly assesses the future significance of its product and demands the full capitalized value of future profits, the price will appear outrageous in relationship to early production and profit levels.

A third possibility is predatory pricing by *z* to eliminate *y* from the market. Predatory pricing in this case, as in general, reduces total industry profits. However it is less harmful to the dominant firm in this case than with undifferentiated products, because the low prices can be confined to *y* and do not need to be offered for the main market of product *z*. If *y* has perfect knowledge and perfect financing, no benefits accrue to *z* through predatory pricing; *y* will simply remain out of the market until *z* tires of reduced profits and then reenter. However, predatory pricing is attractive to *z* if *y* lacks knowledge of *z*'s costs or lacks financing. If *z* can offer *y* products at or below the cost of the *y* firm, it may be able to convince the *y* firm that it has no cost advantage and thus ought to sell out at a low price. If the *y* firm depends on initial profitability to attract financing, the elimination of initial *y* profits through predatory pricing may cause it to accept a reduced price in a merger.

To undertake predatory pricing, *z* could initially produce units of *y* at a cost of $10c_z$ (twice the cost of the *y* firm because of the *y* firm's cost advantages) and sell them at a price of $5c_z$ to deprive *y* of profits. The action would cost *z* actual losses of the units of *y* sold and reduced profits in the production of *z* because of the movement of *z* customers to *y*. Assuming that *z* produced all of the *y* units demanded at the initial cost levels, predatory pricing would reduce *z*'s initial profits to 30 percent below the level with independent pricing. The initial level of predatory pricing profits for z would be the same as its profits eleven years into the competition with independent pricing. The cost of continuing predatory pricing would increase as the cost of producing y declined. Thus z would have to expect a relatively quick collapse of y to make the predatory pricing option profitable.

Case 3: Interdependent Barriers to Entry

This case is a stylized representation of the initial telegraph-telephone competition. Assume that z is controlled by a monopolist which also has patents on y technology. A new firm has additional patents on y technology. Because of conflicts among the patents of the z monopolist and the new firm, litigation between the firms will result in invalidation of both sets of patents and free entry into y. The combination of both sets of patents will produce a total barrier to entry in y. Both the monopolist of z and the new firm initially enter product y.

From z's point of view, control of y along with z is the optimal situation. However, z would prefer to see product y controlled by another monopolist than opened to competition. If y becomes a competitive market, z will not be able to make any profits from sales of y and will also lose demand from z to y. If y is controlled by another monopolist, z will lose potential monopoly profits from y and eventually lose control of the entire industry as y's cost declines, but the displacement will be slowed relative to competition by the monopoly prices charged for y. Thus z would be willing to exit from product y rather than pursue patent litigation to the point of invalidation of the patents.

From the new firm's point of view, a share of a duopoly with z is preferable to full competition. However, full control of y would be superior to sharing y with z. The new firm's bargaining strength is increased by the fact that patent litigation would be more damaging to z than to itself. The least risk strategy is to avoid patent litigation and attempt to work out an agreement for sharing the y market. However, if the new firm is confident that z will not allow full competition in product y, it will file a patent suit against z. If the new firm can make a credible threat to pursue the patent suit to completion, z will find it worthwhile to leave product y and turn its patents over to the new firm in exchange for a promise to maintain monopoly level or higher prices in product y.

3 The Telegraph in the United States

The Invention of the Telegraph

The first telecommunications systems were developed for military use. The most extensive of the early systems was built in France in the early nineteenth century to provide necessary information for the continuation of the Napoleonic wars. The French system consisted of towers erected at intervals of several miles, each equipped with a visual signaling device and a telescope. The operator watched the adjoining towers through the telescope for codes and then relayed the codes with his own signaling device. Although the system allowed communication over long distances with far greater speed than messages could be moved by horseback or ship, the system was cumbersome and expensive to maintain. A similar system was proposed for the United States during the War of 1812, but it was never built.

Following the development of an effective battery around 1800, several experimenters began work on an electric telegraph. Many models were built by scientists of the period, but no attempt was made to commercialize the work. In 1832 the artist Samuel Morse became intensely interested in the possibilities of an electric telegraph after hearing a discussion of recent advances in electricity from a fellow passenger on a ship while returning from a European art study trip. Morse began experimenting with a variety of ideas for communicating via electricity and by 1836 had developed a working model of a simple telegraph.

Morse's device was based on an electromagnet which attracted a lever when the circuit was completed by a distant operator. It was much simpler than models built by the established scientific figures and required the letters to be encoded and decoded by the operators. However, the simplicity also gave the Morse device commercial advantages because of its ease of operation with less than ideal wires and electric current control. Morse engaged the help of two other men to improve his device in exchange for a share in the patent rights, and in 1838 filed a patent application for the Morse telegraph.

Because Morse and his associates lacked the personal financial resources necessary for continued development of the telegraph, and because prior visual telegraphs had been built by governments for military use, Morse sought government funds for an experimental line. Morse's 1838 demonstration of his apparatus was convincing enough to cause the House Committee on Commerce to recommend a $30,000 appropriation to build an experimental telegraph line, but the bill failed to pass the entire Congress. However, the committee chairman, F. O. J. Smith of Maine, was so impressed by the invention that he offered to provide legal counsel and promotional funds to the Morse partnership in exchange for a one-quarter interest in the patent. The offer was promptly accepted. Smith's promotional funds were used to finance a European trip for Morse in an unsuccessful attempt to secure foreign rights to the telegraph. For the next five years, little was done with the telegraph. After renewed lobbying by Morse and his associates, Congress appropriated $30,000 to build an experimental line between Baltimore and Washington. The project was completed in May 1844. The Morse invention established practically instantaneous communication between the two cities. The telegraph received a great deal of favorable publicity when the Democratic National Convention met in Baltimore just as the line was completed and Morse was able to provide communication between the convention and Washington.

For a year the Baltimore–Washington telegraph was operated by the government on an experimental basis without charge. Although it was shown to be a technical success, little demand materialized. In April 1845 the line was transferred to the Post Office and operated under a tariff of 1¢ for four words. It was not a financial success. For the first six months of public operation, total income was $413.44 while total expenses were $3,284.17. The postmaster general argued for government control of the telegraph but did not believe that its revenues would cover expenses under any feasible set of tariffs.[1]

Economic Characteristics of the Telegraph Industry in 1845

In 1845 the telegraph "industry" consisted of the Morse patent and the Baltimore–Washington line, which was not fully occupied even when service was offered at a zero price. Yet from the point of view of the patent holders, the economic characteristics of the industry at that point in time had to be analyzed to determine a rational strategy for the exploitation of the patent. The technological characteristics of the industry and the natural barriers to entry, as well as the opportunities for developing barriers to entry, can be known or estimated before the industry enters the commercial stage. Profit-maximizing entrepreneurs will evaluate those possibilities for a nonexistent industry that they hope to develop just as they will evaluate those possibilities for a mature industry which they hope to enter. Although expectations may turn out to be wrong, it is the entrepreneurs' initial expectations about the economic characteristics of a new industry that determine their initial strategy choices on such matters as short-run profit maximization versus the development of a large market, investing in barriers to entry versus accepting existing levels of entry barriers, or entering the business versus abandoning the enterprise.

The first variable to estimate was demand for the new service. Telegraph service and mail fit into the "industry boundary" model at the end of Chapter 2. Mail was the established service, and it provided communications to customers with a wide variety of needs. Some customers who would have been willing to pay a large premium for greater speed than the mail provided purchased mail service at the normal rates and speed because it was the only communication medium available. The telegraph would provide a much faster and much higher cost means of communication than the mail. It would displace the mail for a specialized segment of customers. The demand for telegraph service would be highly dependent on the characteristics of mail, but the demand for mail service would have little dependence on the prices and characteristics of telegraph service so long as it was priced at a substantial multiple of the mail price. If technological progress should bring the cost of telegraph service down substantially, it could become an important competitor to the mail. The problem facing the patentees was to determine the distribution of consumer preferences between telegraph and mail characteristics and to estimate how many customers would find the telegraph speed adequate compensation for its high price.

If the postmaster general's argument that telegraph demand was so low that no price would yield a profit was correct, the patentees could dispense with any further business analysis and abandon the project or sell the patent to the government at any price they could obtain. However, the patentees had good reason to be skeptical of the postmaster general's pessimistic forecast. It was in his self-interest to encourage pessimistic thinking about telegraph prospects. The mail monopoly would be strengthened by Post Office control of the telegraph, which would be easier to obtain if the investing public had low expectations of telegraph prospects. If Post Office control could not be obtained, the next best outcome was to avoid extensive commercial telegraph development and see the invention remain on an experimental basis.

Several positive signs of profitable demand for telegraph service existed. There was extensive discussion of building a New Orleans–New York visual telegraph for military communications and of building several smaller visual systems for private communications. The electric telegraph was clearly superior to the visual telegraph because it could be operated with equal or greater speed, greater reliability, and lower cost. Insofar as the visual telegraph proposals were serious, they represented demand for electric telegraph. News services and stockbrokers incurred considerable expenses for special couriers, carrier pigeons, and other means to provide communication faster than mail, suggesting potential demand for telegraph service.

The money-losing Washington–Baltimore line was not a good indicator of telegraph potential because the limited distance between the two cities and the existence of train service there limited the telegraph time advantage to a few hours. However, the large distances in the United States and the very limited railroad network in 1845 caused communication to be very slow between many points. Telegraph demand could be expected to be much greater on routes where it could save days or weeks instead of hours when compared with mail service. A further limitation of the Washington–Baltimore line was that it was the only telegraph line in existence. Communications passing through those cities but not originating in one and ending in the other (such as Philadelphia–Washington communications) could only be sent by telegraph if they were relayed by mail where the line did not exist. The time lost in covering some distance by mail greatly reduced the advantage of telegraph communication. The ability to provide telegraph service to other cities would increase the demand for the Washington–Baltimore line by allowing it to be used as a link in communication over longer routes. Thus while demand for the service was uncertain in 1845, there was reason to believe that profitable commercial operation was possible if a fairly complete network could be built to offer service between many widely separated cities.

The next question that faced the Morse group was whether or not enough barriers to entry existed or could be erected to allow them to recoup start-up losses on the enterprise. The most obvious barrier to the entry of others was the Morse patent. The patent did provide substantial protection both in the expectations of the patent holders and the actual events of the industry. Yet the Morse patent, as with most patents, did not provide total protection in itself. It was subject to legal challenges and possible invalidation or narrow interpretation if others found it profitable to mount objections to the patent. It covered only the actual instruments used in sending and receiving the electrical signals, not the wires and poles which constituted the primary investment in an operational telegraph system. A working telegraph using instruments quite different from the Morse instruments already existed in England, and various inventors had developed experimental models of telegraph instruments which were at least potentially usable and noninfringing. Thus the Morse group had to be concerned about other barriers to entry if the industry should be profitable enough to attract significant interest in entry.

The second source of barriers to entry was economies of scale. Economies of scale in the telegraph industry can be divided into technical economies of scale and system economies of scale. The technical economies of scale were quite modest. The two sources of technical economies of scale were the indivisible capacity of a single wire and the ability to string several wires on one pole. Each message occupied one wire for the amount of time required to transmit it. A very low capacity telegraph would be built with a line of poles containing one wire, while higher capacity telegraphs would add wires to the same set of poles. While this effect gave the larger firm a cost advantage over the smaller firm, it was moderated by the fact that the capacity of a single wire was quite low, timber was abundant and poles inexpensive in the mid-nineteenth century, and low-quality telegraphs could be built rapidly by attaching the wire to trees or fence posts.

System economies of scale provided greater potential for protection than technical economies. A company with only a single route was restricted to carrying messages along that route unless it had connecting arrangements with other carriers. If a dominant firm chose to refuse connection privileges to competitors, the potential competitors would be forced to build a complete system in order to compete effectively. If financing, technical economies, or right-of-way limitations prevented the entrant from establishing a complete system, the established firm would benefit from system economies. As discussed in Chapter 1, systems effects can extend other barriers to entry but cannot create barriers if no others exist.

The third potential barrier to entry was from control of the rights of way to string wires. While this would have been a difficult barrier to develop in the beginning, it offered great possibilities as the industry became more settled. Rights of way could be obtained via public franchises along roads and other public lands, via private grants for building on private land, or via arrangements with existing right of way companies such as railroads. In all cases, the firm could seek to obtain rights of way for its own wires in the least expensive way or it could expend somewhat greater effort and money and seek exclusive agreements that would prevent a competitor from using the same right of way. The advantage of seeking exclusive arrangements depended on the alternatives open to another entrant. In the early days of telegraph development, lines were erected largely along public roads and there was little advantage in seeking exclusive rights because of the large number of alternative routes. Later, as the railroads spread and the telegraph lines were largely put up along the same routes as the tracks, it became profitable for the firms to invest in exclusive arrangements because of the limited alternatives available to competitors.

The fourth potential barrier to entry was the capital requirement. A substantial amount of capital would be required to erect a complete telegraph system, and it could be difficult for a new entrant to raise that capital in the relatively fragmented and unorganized capital markets of the mid-nineteenth century. However this was also a problem for the Morse patentees, as they had little personal wealth and their invention was a risky investment.

Given the economic conditions facing the telegraph industry in 1845 (uncertain demand, customers drawn from the Post Office, partial patent protection, need to develop an extensive system, expected difficulties in raising capital), the optimal strategy, from both the telegraph patentees' and the Post Office's points of view, was a purchase of the patent rights by the Post Office. The Post Office had an interest in maintaining a high telegraph price and limiting the incursion of telegraph service on mail according to the "industry boundary" model of Chapter 2. It would thus prefer to have telegraph service under its control or to see full monopoly pricing in telegraph service than to have telegraph become a competitive industry. A Post Office purchase of the Morse patent would allow it to control telegraph development and supplement patent protection with its legal monopoly on mail service. The Post Office could pay the capitalized value of the expected monopoly profits for the patent, operate the telegraph as a monopoly and earn back the money while protecting its mail service from a competitive telegraph industry. From the Morse patentees' point of view, a sale to the Post Office for a suitably high price would be superior to the uncertain prospect of raising capital, developing the industry, and preventing competitive entry.

If a merger with the Post Office could not be arranged, the second potential strategy was to set up a corporation to undertake the unified development of the telegraph system. To ensure adequate demand, an extensive system would have to be built relatively quickly, but building such a system accentuated the problems of raising capital and increased the risk of poor construction. Electricity was not well understood at the time and no one had constructed a system to carry current through hundreds of miles of wire, so considerable experimentation was necessary to determine wire sizes, insulation, and power requirements for a commercial telegraph system. If small experimental systems were set up, they could be expected to lose money initially, causing greater difficulties in raising capital for expansion. Because the patentees lacked personal funds and business experience, the prospect of establishing a company to build a nationwide system of telegraph was not very bright.

The third option open to the patentees was to grant licenses or franchises to local entrepreneurs in various geographical areas and set up a system of independent geographical monopolies coordinated by common dependence on each other and dependence on the Morse patent rights. This option would place the burden of construction and fund raising on local businessmen and greatly reduce the entrepreneurial talents required of the patentees. The main disadvantage of the franchise method would be that it would be difficult to manage the enterprise in a unified way. Cooperation could be induced through the common interest in achieving through routes and protecting the patent monopoly, but it would be difficult to invest in barriers to entry or develop a unified response to a potential entrant.

From Monopoly to Competition, 1845–1851

The Morse interests simultaneously pursued the strategy of sale to the government and the strategy of private development, using the threat of higher prices after the system was developed privately to encourage the government to act. Morse first proposed selling partial patent rights to the government in the summer of 1844. At the same time, public demonstrations of the telegraph were given in Boston and New York in an attempt to raise capital, with little success. Following further unsuccessful attempts to sell the patent rights to the government in late 1844, three of the four Morse patent holders agreed to appoint Amos Kendall, the postmaster general under President Jackson and a man with executive talent, connections, and enthusiasm for the telegraph, as their agent for the exercise of the patent rights. However, Congressman F. O. J. Smith refused to let Kendall represent his 25 percent interest, and according to the terms of the partnership agreement this refusal made agreement between Kendall and Smith necessary to take action.

Kendall pursued the strategy of licensing small geographical companies while holding open the option of selling out to the government. Four separate companies were established during 1845, one to build from Philadelphia to New York, one from New York to Boston, one from New York to Buffalo, and one from Philadelphia to St. Louis. Each of the first three companies assigned 50 percent of its stock to the patent association in exchange for the Morse patent rights. The Philadelphia to St. Louis company was only required to turn over 25 percent of its stock, presumably because of the relative sparseness of the Western population and lower likelihood of substantial monopoly profits than on the East Coast. Each of the 1845 contracts preserved the patentees' option of selling to the government by requiring the new companies to sell their lines to the government under specified conditions.

The amount of capital required for the individual lines was modest even by the standards of that day. The Philadelphia–New York line was built for $15,000, the New York–Boston line for $40,000, and the New York–Buffalo line for $100,000. Most of the capital was raised in small amounts from people located along the telegraph line. Although those who provided capital presumably expected to make profits on their capital, they could also gain part of the return on their money from the benefit of having their town in telegraphic communication with the major cities of the country. Less certainty regarding the ultimate profitability of the telegraph was necessary to induce a businessman who expected to use the telegraph to invest a small amount of capital to build lines through his town than to induce a wealthy capitalist to finance a large part of a telegraph system purely for the potential profits from the investment.

The early companies issued stock with a par value of from four to six times the construction cost of the telegraph line. Part of the difference was used to issue stock of par value equal to two or more times the actual value paid in to the providers of capital, part to issue stock in exchange for the patent rights, and part as promoter's profits for the promoters of the individual lines. In addition to promotional stock, the individual entrepreneurs often made initial profits through awarding contracts for the actual construction of the lines to a construction company owned by the telegraph promoter. The promoter then would receive substantial construction profits in cash out of the original capital of the telegraph company and would find the venture profitable regardless of the ultimate success of the telegraph company. The opportunities for immediate profits as well as potential long-term profits eased Kendall's job of recruiting company promoters, while the ability to raise capital in small amounts all along the line eased the funding of the lines, and telegraph building became a boom industry.

While construction of the initial lines was proceeding, the patentees continued their efforts to convince the government to buy them out. The postmaster general supported nationalization of the telegraph and wrote in his December 1, 1845, annual report:

> It becomes, then a question of great importance, how far the government will allow individuals to divide with it the business of transmitting intelligence—an important duty confided to it by the Constitution, necessarily and properly exclusive? Or will it purchase the telegraph, and conduct its operations for the benefit of the public? . . .

> The use of an instrument so powerful for good or evil cannot with safety to the people be left in the hands of private individuals uncontrolled by law.[2]

However, neither the patentees nor the postmaster general could arouse enough interest in Congress to obtain an appropriation for the purchase of telegraph rights. With the onset of the War with Mexico (May 1846), the patentees tried for government subsidy as a war measure. Kendall proposed that the government build a line from New Orleans to Washington for war use. The patentees would use the line for commercial purposes when it was not occupied with government business in exchange for the government use of the patent rights on the line. Although considerable interest was shown in the Kendall plan, no money was appropriated for a government telegraph line. The patentees then abandoned their efforts to obtain government takeover or subsidy, purchased the Baltimore–Washington line from the government, removed the government takeover clauses from the operating company contracts, and focused their efforts on private development of the telegraph.

In the initial allocation of franchises during 1845, little attention was given to potential conflict of interest situations. The Philadelphia–New York contract was given to Kendall, the trustee for 75 percent of the patent interest, while the New York–Boston contract was given to Smith, the holder of 25 percent of the patent rights. Both men thus assumed the dual role of head of an operating company with a large personal stake in the success of that operating company and holder of patent rights with concern for the promotion of the Morse telegraph industry as a whole. The multiple roles and absence of unified control of the Morse patent rights limited the ability of the patentees to respond to potential competition.

Commercial telegraph operations were begun in January 1846 with the opening of the Philadelphia–New York line. The telegraph line itself only extended from Philadelphia to Newark as the company was unable to build it across the Hudson. Messages were carried by ferryboat across the river and telegraphed the rest of the way. Early business came from stockbrokers, who used the line for arbitrage between New York and Philadelphia stock exchanges, gamblers, and reporters. In spite of crude construction and repair methods and the absence of connecting lines, the Philadelphia–New Jersey line produced enough revenue to cover operating expenses and a modest return on invested capital during the first five months of operation.[3] Profitable initial operation of the New York–Boston line (opened June 1846) and the New York–Buffalo line (opened September 1846) indicated that demand was adequate to produce extensive profits for a coordinted large-scale monopoly telegraph system. The New York–Buffalo line was the most profitable of the 1846 lines, confirming the theoretical expectation that a longer line could compete with the Post Office far more effectively than a shorter line.

The contract for the Western lines (from Philadelphia to St. Louis, with ambiguous authorization for other areas) was awarded to Henry O'Reilly, a flamboyant Philadelphia newspaperman. The O'Reilly contract required him to complete the Western line between a connection with the Philadelphia–New York line and Harrisburg, Pennsylvania, within six months or forfeit the contract. While both the Western line and the Eastern lines were under construction during late 1845, the route of the New York–Philadelphia line was changed to bypass Lancaster, the original connecting point with the Western line. O'Reilly completed his line between Lancaster and Harrisburg within the required six months, but did not make actual connections with the New York–Philadelphia line, raising a question of possible default on the contract.

The initial operating results of the Eastern lines convinced the patentees that they had been too generous in the award of the Western lines to O'Reilly with only a 25 percent stock payment instead of the 50 percent required on the Eastern lines. In spite of low population density, the Western lines were likely to be very profitable and the ambiguous O'Reilly contract could have been interpreted to have awarded the entire territory between the Great Lakes, the Ohio River, and the Mississippi River to O'Reilly. Using O'Reilly's failure to connect with the New York–Philadelphia line in January 1846 as a bargaining chip, the patentees began negotiating with O'Reilly for an increased share of the stock in his company.

O'Reilly refused additional payment and continued building his lines. Unable to secure compliance from O'Reilly, the patentees faced an important strategic question: should they void the O'Reilly contract and risk competition or accept the 25 percent of the stock in his company and maintain Morse monopoly with a substantial share of the monopoly profits going to O'Reilly? Whether or not it was in the interests of the patentees to void the O'Reilly contract depended on three facts, all of which were uncertain at the time of the decision: (1) Would the courts clearly uphold their right to abrogate the O'Reilly contract based on his default? (2) Would the courts clearly uphold the validity of the Morse patent? (3) Would anyone be able to develop a telegraph instrument in the near future that would not violate the Morse patent?

It would only create disharmony and bad working relationships to attempt to void the contract and then have the validity of the contract upheld by the courts. If the patentees were successful in changing O'Reilly's status as a licensed user of the Morse patent, he could be expected to turn his energies toward invalidating the patent, either by a direct court challenge to its validity or by an attempt to find noninfringing apparatus. So long as O'Reilly was a Morse licensee, he had an incentive to support the validity of the Morse patent against any potential challengers, but if his license was revoked, he would have a strong incentive to challenge the Morse patent and seek out other inventors in order to protect his investment in telegraph lines. While it was unlikely that O'Reilly would have sought entry without a license from the Morse patentees in 1846, it was also unlikely that he would quietly exit if his license was revoked. If O'Reilly could find a way to operate without the Morse license, he would be a formidable competitor to the Morse licensees because he was nearing completion of the Pittsburgh–Philadelphia line and had as much access to capital and expertise as the patentees.

If the patentees could be certain that operation without Morse patent rights was impossible, they could transfer profits from O'Reilly to themselves by breaking the contract. The Morse interests had to balance the benefits of a higher percentage of profits from the O'Reilly territory against the potential loss of monopoly profits if breaking the contract caused entry outside of Morse patent control. If O'Reilly and the patentees agreed on the significance of the Morse patent, agreement could have been reached on the amount of royalty payments. O'Reilly would increase payments rather than be forced out of business if he believed the patent was a total bar to new entry. The patentees would accept 25 percent of O'Reilly's stock rather than nothing if they believed he could operate without their authorization. The uncertainty of the situation allowed the parties to have different expectations and thus fail to come to agreement if they were not sufficiently risk averse.

The risks and potential rewards of voiding the O'Reilly contract caused a split among the patentees, who differed in their levels of risk aversion and their expectations for the future. Kendall advocated the less risky course of maintaining the O'Reilly contract and accepting a smaller share of monopoly profits. In October 1846 he wrote to Smith: "If we manage with prudence and act in concert, the revenues of a nation are within our reach. If we are divided in counsel or raise the public against us, we jeopardize everything and shall live in constant turmoil."[4] Smith was more confident of the strength of the patentees' position and advocated voiding the O'Reilly contract and awarding a new contract requiring 50 percent stock payment for the patent license.

The organization of the patentees, as a partnership that required agreement between Kendall and Smith for action rather than as a corporation, gave the early patent monopoly some of the characteristics of a cartel. While Smith and Kendall were in agreement that a strong patent monopoly with a large share of the operating company stock awarded to the patentees was in their interest, their differing expectations for the future and differing personal stakes in the various possible outcomes made it difficult to agree on actions that would have been industry profit maximizing. If the patentees were going to undergo the risk of competition from O'Reilly, it would have been in their interest to take compensating actions, such as investment in the experimental telegraph apparatus of other inventors, to strengthen their patent position. However, as Caves and Porter have suggested, barriers to entry can be thought of as a collective social good for the industry.[5] A dominant firm or monopolist finds it profitable to invest in barriers to entry, but because the barriers benefit everyone in the industry regardless of which firm pays for them, there is an incentive for firms in a cartel to underinvest in barriers to entry in the hope that other firms will pay the cost of developing the barriers. Thus it may be difficult for collusive firms to agree on investments in barriers to entry. In the telegraph case, the partnership could not agree on barrier-creating expenditures, either through allowing the O'Reilly contract to continue (an action which in itself could have been considered an investment in barriers to entry, because it meant lower immediate profits with a reduced risk of later competition) or by purchasing the rights to other telegraphic inventions.

Because Smith was counsel to the patentees and the critical factors in the decision were legal questions of the validity of the O'Reilly contract and of the Morse patent, the patentees followed Smith's advice to void the contract. In November 1846 the patentees gave formal notice to O'Reilly that his contract was null and void. They then began to allocate the O'Reilly territory to new promoters under contracts that called for 50 percent of the stock of the operating company to go to the patentees. The new contracts called for rapid building of lines from Buffalo to Milwaukee, Cleveland, Cincinnati, Louisville, and other cities. The plan was to isolate the O'Reilly line by connecting the western cities with the East Coast via the New York–Buffalo line rather than O'Reilly's Philadelphia–Pittsburgh line. This plan allowed the patentees to attempt to block operation of the O'Reilly line through legal action without cutting off connections with the West or requiring the building of a new Philadelphia–Pittsburgh line while there was a possibility of acquiring the O'Reilly line at a distress price.

When the contract was declared void by the patentees in November 1846 O'Reilly had almost completed the Philadelphia–Pittsburgh line. He continued his building program, and maintained that the contract was still in force. He held 25 percent of the stock of his company available for the patentees (who refused to accept it) and asserted his status as the only legitimately licensed telegraph promoter in his area to the public and potential investors. However, rather than depend on the status of his contract with the patentees, he purchased partial rights to the printing telegraph of Royal House, which was in the development stage at that time with a patent pending. The House instrument was a significant technological advance because it printed letters directly upon receipt rather than the dots and dashes of the Morse receivers. It was faster and required less skillful operators than the Morse instrument but was a complex instrument which was subject to breakdowns and needed higher quality electrical signals than the Morse instrument to operate properly. O'Reilly investigated other telegraph inventions and secured some rights to them. In addition O'Reilly increased the speed of his building program in order to develop an extensive network of lines under his control. While building new lines was very risky given the possibility of being legally prevented from using them, the building program reduced O'Reilly's dependence on connections with the Morse lines and gave him a strong competitive position if the patent barriers could be broken down.

When O'Reilly opened the Philadelphia–Pittsburgh line in January 1847, the patentees filed for an injunction to stop commercial operation. The injunction was not granted. Three different suits were filed that year against O'Reilly in different jurisdictions, but the patentees were not able to obtain a decision that would immediately block the operation of the O'Reilly line. With the legal situation unsettled, the patentees decided to expand the lines under their control as rapidly as possible and also declared a policy of noninterconnection with the O'Reilly lines. A noninterconnection policy is a superior competitive weapon to a price war for a dominant firm because it has the greatest effect on the smallest firm. However, the noninterconnection policy does impose costs even on the dominant firm. Because the telegraph operating companies were not under the unified control of the patentees, not all were willing to bear the costs of isolating the O'Reilly system and the noninterconnection policy was imperfectly enforced.

With the spur of competition, telegraph construction activity was intense during 1847. O'Reilly's Philadelphia–Pittsburgh line, which opened in January of that year, was extended to Cincinnati in August, to Louisville in September, and to St. Louis in December, giving O'Reilly telegraph communication from the East Coast to the western edge of the country. O'Reilly also built spur lines to his main line and began construction along the Great Lakes during that year. O'Reilly's Great Lakes construction was in direct competition with the construction of the company to whom the O'Reilly territory was licensed by the patentees. Because both companies depended on local investors along the route for capital, and both claimed to be the only legitimate Morse license holder, public uncertainty over the outcome of the legal dispute restricted investment and slowed construction activity. Each company waged a propaganda campaign against its opponent.

Encouraged by the profitable operation of his initial lines and the hero's reception he received in western towns, and angered by the refusal of connection privileges with other lines and the direct competition of the patentees in his territory, O'Reilly abandoned the guise of a Morse licensee and determined to build a nationwide system with non-Morse instruments. In the fall of 1847, he withdrew his offer to pay 25 percent of the stock in his companies to the Morse patentees and began building south from Louisville toward New Orleans, an area outside of his original patent grant. As in the Great Lakes area, the Morse licensed company engaged in a construction race with the O'Reilly company along the Louisville–New Orleans route. The Morse contractor for the Louisville–Nashville section described the competition as follows:

> For 15 miles they were side by side and when a man finished his hole he ran with all his might to get ahead, but finally on the 24th, we passed them about 80 miles from here and now we are about 25 miles ahead of them without the loss of a drop of blood, and we shall be

able to beat them to Nashville if we can get the wire in time, which is doubtful.[6]

O'Reilly's decision to withdraw his offer of stock to the Morse patentees and to build outside of his original territory made the pending legal question of whether or not he had forfeited his contract moot and caused him to be dependent on finding a noninfringing telegraph apparatus. He had decided that the House apparatus was too intricate and unreliable, and had given up his rights to the House invention. O'Reilly first turned to an imitation of the Morse device known as the Columbian telegraph. The Morse group filed patent infringement charges and in September 1848 succeeded in obtaining an injunction in Kentucky against the use of the Columbian telegraph. Because only the sending and receiving instruments violated the Morse patent and not the telegraph lines themselves, O'Reilly hoped to escape the Kentucky injunction by having only lines and not telegraph instruments in Kentucky. Consequently he moved the Louisville office across the river to Jeffersonville, Indiana and transmitted messages by hand into and out of Louisville. The Kentucky authorities were not impressed with O'Reilly's ingenuity and the marshal seized the O'Reilly lines. Because Louisville had been the major switching point for O'Reilly's system, the Kentucky seizure split the O'Reilly system into isolated local lines without the ability to carry long-distance transmissions.

Following the failure of the Columbian telegraph to stand up to court challenge, O'Reilly purchased the United States rights to the telegraph apparatus of Alexander Bain, which had been developed in Scotland and patented in England. The Bain telegraph used chemical action to record the electrical signals rather than an electromagnet as in the Morse device. O'Reilly agreed to help Bain secure a United States patent on his device and continued building his line toward New Orleans, while planning a Bain telegraph system along the East Coast in direct competition with the Morse lines there.

The O'Reilly line to New Orleans was completed in January 1849 but was of limited usefulness because of the seizure of the Kentucky lines. However, in March of that year, the Bain patent was granted giving O'Reilly a presumption of legality in proceeding without a Morse license. The Morse patentees continued to assert that the Bain apparatus infringed the Morse patent and filed suit against O'Reilly but did not meet with immediate success. Following the granting of the Bain patent, the Kentucky lines were returned to O'Reilly. Financial and managerial problems delayed the completion of the Morse line to New Orleans until early 1851, but the patent litigation and natural disasters prevented O'Reilly's lines from making extensive profits during the two-year period of monopoly. O'Reilly's Kentucky lines had deteriorated significantly while under the control of the marshal and required extensive rebuilding after their return. A Mississippi River flood during the spring of 1850 washed out large portions of the line and required further rebuilding, leaving the O'Reilly New Orleans line with only a short time of effective service before the Morse line was completed.[7]

While the Morse patentees were busy challenging O'Reilly's Columbian and Bain telegraphs during 1848, they encountered further competition from the improving House printing telegraph. Following O'Reilly's rejection of House patent rights, the rights to the device were sold to various other potential telegraph entrants. The first operational House line was a Philadelphia–New York link completed in April 1848. As underwater insulation technology was still undeveloped at that time, the House company used a 350-foot-high mast to carry the line across the Hudson River and provide direct telegraphic communication with New York. This tactic gave the House line a speed advantage over the Morse line, which had continued to carry messages across the Hudson by ferryboat for its two-year monopoly period. The Morse patentees filed a patent infringement suit against the House company, but were unable to get an immediate injunction to stop the operation of the House line. The House company continued to extend its system and during 1849 completed the Boston–Washington and the New York–Buffalo links, providing direct route competition to all three initial Morse routes.

Armed with the Bain patent rights, O'Reilly also built competitive Boston–Washington and New York–Buffalo lines during 1849. The Eastern lines gave O'Reilly a nationwide system with the ability to transmit messages from Boston to New Orleans. Although all of the lines in the O'Reilly system had been promoted by him, his system was not unified under his control. Each significant line had been set up as a separate company under the control of the local capitalists who financed the line. O'Reilly received substantial stock in the various companies for his promotional efforts, but he did not hold a controlling interest in each one. O'Reilly's personal control over the lines was further diluted by the rapid expansion of the system and the legal inability to operate some of the lines which produced a liquidity crisis for O'Reilly and caused him to turn over his financial affairs to banking trustees to avoid bankruptcy. Some of the individual lines were forced into reorganizations by creditors. Thus the O'Reilly empire was limited in its ability to mount a coordinated attack on the Morse and House lines when significant competition among the operating companies began.

With three competitive companies on the Boston–Washington and New York–Buffalo routes during 1849, the first serious price competition of the industry occurred. The Boston–New York rate of 50¢ for ten words dropped to a minimum of 1¢ per word at the height of the rate war. A similar rate war occurred on the New York–Buffalo line. Although dividends were eliminated or cut to nominal levels during the period of heavy competition, there is no evidence that the competitive rates were below operating costs. All three lines continued operations until 1851 when the Bain patent was found to be an infringement of the Morse patent.[8]

By the beginning of 1850, the telegraph industry had lost its monopoly characteristics. No major routes were without competition. Companies operated under the Morse, Bain, or House patents and in some cases without any patent license. Although the system economies of scale and the patent infringement litigation made potential barriers to entry high, the actual barriers to entry at the time were almost nonexistent. Usable systems could be built very quickly by stringing wires on brackets nailed to trees as well as by high-quality construction. The fact that actual patents had been granted to both Bain and House gave non-Morse apparatus a presumption of validity that was sufficient to encourage entrepreneurs to use them even though the Morse interests were filing infringement suits.

Just four years earlier, the industry could be seen as a unified cartel with some barriers to entry. Even though the lines were just being built, all of those under construction were licensed to the Morse patent and had granted from 25 to 50 percent of their stock to the patentees in exchange for the privilege of using the Morse patents. If the Morse patents had been controlled by one person or a single company, it would have been clearly advantageous to extend control via that patent and to develop barriers to the entry of new firms. It would have been profitable to purchase control of the various new inventions that could substitute for the Morse device. At the time they could have been purchased for relatively small amounts and used to create barriers to entry. However, the division among the holders of the Morse patent prevented them from acting in unison to maximize profits. Differing opinions of the risk involved and different levels of willingness to bear risk led to the long, drawn out O'Reilly struggle, which led to direct competition and provided a market for non-Morse telegraph equipment. The existence of already constructed O'Reilly lines in need of only an instrument to operate provided a far greater stimulus to invention than the prospect of facing an entrenched monopolist with the need to not only invent an instrument but also to build a system in order to compete.

Overconfidence in the significance of the Morse invention prevented the patentees from purchasing the alternative inventions. This probably came about by the controlling interest being held by Morse himself who could not be entirely objective about the relative merits of his device and others. Instead of making efforts to secure control of other devices, he made every effort to discredit them and show in public writings that they were not real improvements over his own invention. The combination of an overestimate of the degree of protection from competition provided by the Morse patent and divided control over the development of the industry led to a lower than optimal investment in barriers to entry and the rapid loss of monopoly power by the original patentees.

From Competition to Monopoly, 1851–1866

In 1851 the Bain patent was ruled invalid and the Bain device was found to infringe the Morse patent. This event marked the end of O'Reilly's activities as a telegraph promoter and forced his companies to look for a merger partner. The Morse challenge to the House patent had ended unsuccessfully the year before. Consequently the Bain lines could not be forced out of business and purchased for scrap value but were subject to competitive bidding by the Morse and House lines. Because the construction of the Bain lines was more suitable for Morse operation than for House operation, most of the Bain lines were purchased by the Morse line that operated along the same route. The purchase terms varied from line to line but were closer to the terms one would expect of a merger of two viable companies than of the purchase of an enterprise without legal rights to operate. For example, the Bain–Morse consolidated New York–Boston line issued two thirds of its capital in exchange for the three-wire Morse line and one-third of its capital in exchange for the two-wire Bain line, a division almost in proportion to the capacity of the respective lines.

The Bain–Morse mergers eliminated significant competitive activity for a time. They restored a system of local monopolies in the West where House lines had not been built. The House lines on the Boston–Washington and New York–Buffalo routes continued to provide some competition, but the strongest competitor to the Morse lines had been the Bain lines. The House apparatus still was not perfected and could not fully compete with the simpler but more reliable Morse equipment. The local monopoly lines in the West varied in profitability. Many had been built rapidly and carelessly, or built before construction requirements for high-quality telegraph lines were well understood, and required extensive maintenance or rebuilding in order to provide reliable service. Some had been so overcapitalized with watered stock or had been so burdened with high-cost construction contracts (to the promoter's construction company) and promoter's fees, that they were in financial difficulty even though they were making high profits on the physical assets employed. Ownership was divided between local capitalists who were primarily concerned only about the profitability of the local company, and promoters and patent holders who held stock in a large number of local companies and were concerned for the profitability of the telegraph system as a whole. The large number of local companies with no single dominant stockholder to force coordination limited the ability of the telegraph industry to protect its monopoly position. The situation provided a good opportunity for a merger promoter, but the sense of local pride and regionalism was strong enough in pre–Civil War days to cause considerable resistance to any large-scale merger scheme.

In 1851 the New York and Mississippi Valley Printing Telegraph Company was formed and purchased rights to the House telegraph in the West. The company encountered difficulty in raising capital for construction and only built a Buffalo–Louisville line during its first two years. However, in 1854 the company began rapid expansion through a partnership with several railroads. During the 1850s American railroads were beginning to realize the value of having telegraph lines along their tracks. Because many railroads had single tracks, telegraph communication regarding the location of trains was practically a necessity to prevent accidents when trains did not run exactly on schedule. The Mississippi Valley contract with cooperating railroads provided that the railroad would build a telegraph line according to telegraph company specifications, provide suitable telegraph offices along the line, transport telegraph employees and materials without charge, observe the condition of the telegraph lines, and refuse to let any other telegraph lines be constructed along the railroad right of way. In return the telegraph company agreed to issue $125 worth of telegraph stock per mile of line to the railroad, maintain agents at offices along the railroad, maintain the lines in good working order, transmit railroad messages without charge, and give priority to the lines for messages related to the operation of the railroad.[9]

The railroad contracts provided an opportunity for the Mississippi Valley Company to enter the industry in competition with the established Bain–Morse lines. For the most part, the established telegraph lines were not along railroad rights of way because the railroads were very limited in coverage during the first period of telegraph building. To provide railroad service, new lines were needed, either from the established companies, a new company, or the railroads themselves. The railroads' incentive to enter the industry directly was reduced by the need to secure patent rights and the problem of generating commercial traffic over the short distances covered by individual railroads. The Mississippi Valley Company's offer to provide telegraph service for free once the railroad built the line was superior to having a captive line with only railroad messages. If the telegraph company were profitable, the railroad would receive some return on its telegraph stock and have service for less than the cost of building the lines; if not, the railroad could consider its investment in building the lines a prepayment for the telegraph service.

From the Mississippi Valley Company's point of view, the railroad contracts provided a method of quick expansion. By using railroads to finance construction, the company only needed to generate enough commercial business to pay operating expenses in order to be profitable. The absence of construction debt would allow the company to withstand any rate wars without great danger of bankruptcy. The exclusivity provisions of the contract provided a source of barriers to entry if the company should become dominant. Railroad rights of way provided easy access and maintenance, as well as the most useful routes. Thus a railroad-based telegraph line could gain a cost advantage over a road-based line. The established companies did not actively compete with the Mississippi Valley Company for railroad privileges at first. They already had adequate capacity and were not anxious to occupy the railroad rights of way as a form of barrier to entry. The compatibility of railroad and telegraph operations was an idea that was sold to the railroads by the Mississippi Valley railroad experts, not an idea that was recognized as obvious good management at the time. Thus the Mississippi Valley Company was able to make contracts with a number of railroads (but by no means all the lines in existence) before other companies sought similar contracts.

Buoyed by its successful railroad promotions, the ambitious and innovative Mississippi Valley Company made plans for a large-scale telegraph system. The company determined to establish a monopoly system through merger rather than price wars. However, many of the local Morse–Bain companies were resistant to merger overtures. The resistance came from local pride and from potential adverse effects on other independent companies from the merger of one company into the Mississippi Valley Company.

Each company's first preference was to be a local monopoly controlled locally. However, in general the company preferred to be part of a national monopoly rather than to face competition as a local company. This required local stockholders to guess the decisions of connecting companies when faced with a merger proposal. If no other companies would accept merger terms and competitive lines would not be built in the area, the company would resist the proposal even if given a chance to capitalize monopoly profits into its price. If competitive lines would be built, the company would prefer to capitalize the monopoly profits by merger. If connecting companies would merge, the remaining company's value would drop because of reduced connections, even without direct competition in its territory. Thus the first company to accept a merger proposal could get the full monopoly capitalization, while later ones would be forced to take less.

The Mississippi Valley Company thoroughly understood the significance of connecting arrangements in the value of telegraph service. It took advantage of connecting arrangements, lack of unified control of the companies, and the threat of competition to induce mergers with very little actual price competition. The Mississippi Valley Company first sought a group of lines along the Great Lakes, which had been promoted by Ezra Cornell under license from the Morse patentees after the reallocation of the O'Reilly territory in 1847. Cornell held substantial stock in each of the companies, but not controlling interest, and was not managing the lines. Cornell thus had an interest in the performance of a group of lines, while the controlling stockholders were primarily concerned with their individual line. The Mississippi Valley Company approached the local managers of one of the lines and asked them to name a price for the line. A high price was named, but the Mississippi Valley Company accepted in order to gain an entering wedge and Morse patent rights. Because of his interest in other lines, Cornell was strongly opposed to the deal as shown by the following excerpt of a letter from Cornell to the president of a connecting telegraph company:

> Speed sold all the Morse patent, my interest as well as his, thus giving the House folks our own tools with which to defeat us in our operations for extending our lines, and fortifying and protecting our business.
>
> I asked the Colonel mildly why he sold my interest in the patent. His answer was "to make money, by God."
>
> This is the foulest piece of treachery toward me that I have ever known and the reason assigned is appropriate; it is the reason for which Judas betrayed his Lord, and for which Arnold betrayed his country.
>
> How I stand, or what position I occupy, I am unable as yet to understand, as all parties to the foul bargain refuse even to let me see the contract. I am looking about to see what can be done by way of protecting what I have left.
>
> I write this early to you that you may stand on your guard, if any of the conspirators approach you.[10]

Despite Cornell's initial attempt to organize other companies against the merger attempts, he decided that the best "way of protecting what I have left" was to allow the merger of his remaining companies. The acquisition of the Cornell lines was completed in 1855, the year after the initial merger, and the expanded Mississippi Valley Company was renamed Western Union.

The next Western Union target was the National Lines, a group of lines promoted by O'Reilly, with separate ownership but a common superintendent. Western Union gained control of one of the smaller lines by getting its superintendent to lease it from the stockholders and then transfer the lease to Western Union, but it found the major lines resistant to merger overtures. Western Union then undertook a subtle threat to start an opposition line from Louisville to New Orleans, by sending a representative to the South to transmit messages regarding the price of telegraph supplies via the southern telegraph. Superintendent James Reid described the situation as follows:

> Hiram Sibley, of Rochester, now on a general forage for the Western Union Company . . . sent Edward Creighton south, with directions to show himself at certain places and to telegraph fully the price of poles, the best routes for lines, and so forth. Occasionally, a cypher message, which meant nothing, was sent through. Of course these dispatches were not designed to be "strictly confidential." They soon became known at headquarters. Mr. Douglass, as he read these messages, looked solemn. The Doctor's smile was not gay.[11]

The first proposal made to the southern company was merely that it break its exclusive connection arrangement with the National Lines and open connection to the company that offered the largest rebates on connecting business. The southern company agreed to do this in order to avoid the possibility of facing direct competition. Western Union then offered a favorable merger agreement to the Pittsburgh, Cincinnati and Louisville Company, using the threat of deprivation of the connecting lines to overcome local pride. The agreement was accepted and then the southern company was isolated and forced to come to terms. As Reid described it:

> He [Sibley] took measures to convince the Managers of the Pittsburgh, Cincinnati and Louisville line that their southern connection was gone. This was the lemon squeezer which led to the immediate lease of the latter line to the Western Union Telegraph Company . . . This done, Sibley again called at Louisville, but, having now the trump card in his hand, he had nothing to say about rebate. He had bagged both birds and chuckled over his success. He was a royal egotist, and enjoyed his triumph grandly.[12]

The last major link of the National Lines, the Philadelphia–Pittsburgh line, resisted merger attempts. Western Union arranged with the Pennsylvania Railroad to build a line along its right of way to provide competition. The line was built during the summer of 1856 and the competition of the new line combined with its exclusive connections to the west quickly changed the minds of the directors of the existing Pittsburgh–Philadelphia line and a merger was arranged in January 1857. This transaction completed a three-year program which transformed

Western Union from an insignificant company to a major company controlling many of the routes of the West and with favorable railroad contracts and both Morse and House patent rights as protection to its position. The effort was accomplished with very little actual competition, substituting favorable merger arrangements which all parties could see were an advantage to them compared to extensive price wars.

While Western Union was consolidating the western telegraph companies, another new company attempted to consolidate the eastern companies. In 1855 the American Telegraph Company was organized and purchased rights to a new printing telegraph developed by David Hughes. It leased several small lines and offered to consolidate the major Morse lines. Rather than accepting the offer, the Morse lines under Kendall and Smith began their own program of consolidation to strengthen their position against the new entrant.

Failing to gain direct control over the eastern companies, the American Telegraph Company next attempted to set up a formal cartel using the pool of patents as barriers to new entrants. Representatives of all the major companies met in 1857 but failed to reach a unanimous agreement because of a dispute over assessment of the cost of the Hughes patent to companies that did not plan to use the Hughes device. Not all companies agreed on the need for jointly controlling the various patents. Finally an agreement (known as the "Treaty of Six Nations") was reached by all companies except the Washington to Boston lines controlled by Kendall and Smith. The cartel companies agreed to divide the cost of the Hughes patent, restrict operations to allotted territories, refuse patent rights to any company not in the cartel, and provide exclusive connecting arrangements with cartel members.[13] The companies also agreed on a regular schedule of meetings and an arbitration procedure.

The 1857 cartel isolated the Smith–Kendall lines but did not force them into easy submission. The cartel lacked capacity between Washington and Boston and the American Telegraph Company began building along that route. Smith and Kendall responded by extending their lines north toward Portland, west toward Cincinnati, and south toward New Orleans. The next year, the cartel convention defeated a proposal to carry out extensive price warfare against the Smith–Kendall lines and divide the losses among the cartel companies, and instead agreed to seek a negotiated settlement. Smith and Kendall offered to sell their lines at very high prices, which caused dissension in the cartel over payment.

The lines were primarily in American Telegraph's territory, but American Telegraph sought contributions from the other cartel companies in order to avoid potential competition. In 1859 an agreement was reached in which the Smith-Kendall lines and patent rights were sold with the American Telegraph Company paying 50 percent, Western Union 25 percent, and the smaller companies in the cartel the remaining 25 percent. This event established control of the industry under a formal cartel with patent rights to all significant telegraph apparatus in the United States.

In the dispute with the cartel, Kendall and Smith's history as combative risk takers improved their bargaining position. While the negotiated settlement was clearly industry profit maximizing, the division of gains among the individual parties was influenced by expectations regarding their willingness to take losses. If the cartel believed Kendall and Smith would accept modest merger terms immediately upon the initiation of price warfare, it would not have acceded to the price demanded for their property. However, Smith and Kendall's willingness to risk competition in the O'Reilly dispute, and their continuation of the fight through both commercial and legal means after the granting of the Bain patent, as well as their refusal to pay a share of the Hughes patent suggested an independent (and perhaps irrational) spirit which would not easily give in to pressure. The Smith-Kendall lines were well established and profitable, giving them considerable financial strength to bear losses and extend their lines, while the cartel was dependent on agreement of its members to take action. The ability to convince the cartel that they would fight vigorously allowed Smith and Kendall to appropriate a large share of the benefits of the negotiated agreement for themselves. By giving rewards to a company that refused to play by the cartel rules, the agreement also provided encouragement to other companies to violate the cartel agreement when it appeared profitable to do so.

By the end of the 1850s there was considerable interest in a telegraph line to the Pacific. The American company favored a southern route from New Orleans, which would allow the transmission of Pacific business over its lines to New York and Washington, while Western Union favored a northern route via Salt Lake City, which would bring the business through its territory. The original cartel agreement had reserved the rights to a Pacific line as common property of the association. Neither company wanted to build a line without a government subsidy and considerable efforts were made to get a subsidy bill through Congress. The bill introduced by the telegraph interests was substantially modified before passage in June 1860. Among the more significant changes were that the subsidy was to be limited to $40,000 per year rather than the $50,000 requested by the telegraph companies, and the rights were to be awarded by open bidding rather than by naming the telegraph companies in the bill. When the telegraph cartel met in August 1860, it voted against proceeding under the legislation by the association and chose a committee to develop plans for more desirable legislation. Western Union then entered a bid at $40,000, the only bid actually received in the required manner, and was granted the contract.

Western Union carefully organized the project during the early spring of 1861 and started actual construction in July of that year. The work was pushed with great vigor. Despite the logistical difficulties of carrying on a massive construction project prior to the building of the transcontinental railroad, the project was completed in four months. Demand for the line was heavy and the initial rates were set at $1.00 per word, ignoring the provision in the subsidy legislation for maximum rates of $3.00 for ten words. The total cost of the line was approximately $500,000. The $40,000 per year government subsidy provided an 8 percent return on capital if the public business could pay direct operating expenses. In actuality the public business was extremely heavy and profitable and the line's capitalization was accordingly increased. The Nebraska to Salt Lake portion of the line cost $147,000 to build and was eventually capitalized at $6,000,000 worth of Western Union stock.[14]

The Pacific line was the second profitable break with the cartel. While total cartel profits on the Pacific line could have been greater with a larger government subsidy, the legislation passed provided ample opportunities for profits. Western Union's decision to build the line itself after the cartel voted to seek higher subsidies indicated a willingness to risk the disapproval of the cartel to take advantage of a profitable opportunity. As with the Kendall–Smith break, it was a very profitable decision for Western Union. Western Union's right-of-way land and subsidy from the government contract gave it an advantage over any potential competing lines to the Pacific.

While the Pacific line was under construction, the Civil War began. Telegraph demand increased dramatically and the lines of all companies were filled to capacity. Western Union was better situated to take advantage of the war demand because its network was concentrated in the North, while the other major companies had networks that extended into both North and South. Lines between the North and South were cut and companies that extended into both were split into two pieces. Another significant competitive advantage for Western Union was the fact that its general manager, Anson Stager, was commissioned chief of U.S. military telegraphs. He remained general manager of Western Union as well all during the war. Under the direction of Stager, 15,000 miles of military telegraph were constructed during the war, most of which was turned over to Western Union at the close of the war.[15] Western Union enjoyed immense profitability without concern for competition during the war years and expanded its capacity rapidly. The company increased dividends on nominal capitalization (which included capitalization of expected monopoly profits in some cases) from 5 percent in 1861 to 9 percent plus a stock dividend of 27 percent in 1862, 9 percent plus a 100 percent stock dividend in March 1863, another stock dividend of 33 percent in December 1863, and a further stock dividend of 100 percent in 1864, making a sevenfold increase in the nominal value of Western Union stock during the war years.

Although Western Union was relatively unconcerned about competitors during the war, it was not protected from the entry of new firms. The Morse patents expired in 1860 and 1861, leaving a hole in the shield of protective patents gathered by the cartel. A new entrant could be denied access to the printing telegraph apparatus of House or Hughes, but the original Morse apparatus was available to all and still practical for commercial operations. The excess demand and high Western Union profitability during the war years provided strong incentives for new companies to enter the industry. A number of new companies began telegraph service during the war and merged in 1864 to form the United States Telegraph Company. At the time of the merger the new company had 10,000 miles of line and the ability to provide service to many areas without connections to the cartel companies.

In general, the lines of all companies were fully occupied during the war, reducing the incentive and the ability of the cartel companies to take action against the new company. At the end of the war (April 1865), the United States had three major telegraph companies: Western Union with 44,000 miles of wire, American with 23,000 miles, and United States Telegraph with 16,000 miles. The smaller companies in the cartel had either been absorbed through merger or had been so damaged during the war that their operations were of little competitive significance. Demand for telegraph service dropped substantially and

Western Union and American began a double attack on United States Telegraph with a price war and legal challenges to the rights of way used by the new company. The stock of United States Telegraph dropped from near par in early 1865 to 20 percent of par in the fall of 1865. By that time the company was losing $10,000 per month on operations without taking account of capital charges. Because United States Telegraph was the smaller company, had less financial strength, and had inferior lines constructed in haste to meet war requirements, the two major companies could have driven it out of business with a continuation of the price war. However, after a short price war in the fall of 1865, Western Union offered to purchase United States Telegraph for an issue of stock that would give the new company the same proportion of total Western Union stock as the proportion of its operating revenues in the combined revenues of the companies. This was a generous offer given the market price of United States Telegraph stock and the competitive prospects for the company. It was promptly accepted.[16]

The merger of Western Union and United States Telegraph ended the cartel and put American Telegraph on the defensive. Because United States Telegraph had operated in both American and Western Union territory, Western Union gained lines directly competitive with American through the merger. Western Union no longer needed connections with American to provide nationwide service, but American needed Western Union connections. Faced with a choice of merging or building additional lines to compete with Western Union, American chose merger and in June 1866 joined Western Union in return for almost $12 million in Western Union stock. The American merger completed the expansion program begun by Western Union twelve years earlier and established a single monopoly company for the telegraph industry. At that time, Western Union was the largest company in the United States.[17]

Western Union Dominance, 1866–1877

Following the achievement of monopoly in 1866, Western Union had substantial but not perfect barriers to entry. Some protection came from patents, but the expiration of the Morse patent eliminated patents as a significant barrier to entry. The primary barriers were the extensive system possessed by Western Union and the control of favorable routes with exclusive railroad contracts. The large system gave Western Union an advantage over any small competitor. Route control made it difficult for anyone to build a complete competitive system.

Western Union's strategy immediately after the Civil War included high but gradually declining prices, rapid expansion, and technological innovation. During the first decade after establishing its monopoly, Western Union paid dividends of 2 to 8 percent each year on nominal capitalization (which included capitalized expected monopoly profits from the series of mergers, and thus was far greater than the replacement cost of the physical capital), while completely rebuilding and expanding its lines out of current profits. Over the ten-year period, the Western Union offices increased from 2,565 to 7,500, the wire miles from 85,291 to 194,323, and the messages transmitted from 5,879,282 in 1866 to 21,-158,941 in 1876, while the charge per message was decreased 62 percent.[18] The rapid expansion was both a response to profit opportunites and a form of protection from new competitors. The first decade after the Civil War was a time of great expansion of the railroads. If Western Union had not built along the new lines, other companies would have had a natural source of entry into the industry. The rapid expansion could also be seen as a strategy to build ahead of demand and avoid the incentives for entry found by a company such as United States Telegraph during the Civil War.[19]

The high profitability of Western Union attracted the attention of potential entrants, but they encountered considerable difficulties starting up service even without specific opposition from Western Union. Indicative of the problems encountered by new entrants was the experience of the Franklin Telegraph Company. The company was organized in 1865 to exploit a new invention and received a charter to build in Massachusetts. The invention turned out to be less significant than originally hoped and the promoters did not actually build lines. Another enterprise began building a competitive line from Boston to New York without a charter and was denied rights to build in Springfield, Massachusetts. The two companies then merged in order to benefit from the original charter. A third company developed a new invention and received the rights to build through New York City. That company's invention also failed to work properly, but when Franklin Telegraph found it could not get permission to build inside New York City, the two companies were merged to gain the charter rights inside the city. By 1867 the company was in operation from Boston to Washington. The company was unprofitable at first, but began earning profits in 1872, seven years after the original company was founded.[20]

Similar but smaller local competitors entered the industry in a variety of areas. By the time of the 1880 census, there were eighteen railroad telegraph companies and fifty-eight commercial telegraph companies besides Western Union, but Western Union carried 92 percent of the messages and received 89 percent of the total telegraph revenue.[21] Thus entry occurred under the Western Union pricing policy, but at a slow rate which did not threaten Western Union's dominance of the industry.

Western Union faced potential challenges to its monopoly power from two sources: (1) a serious effort by a major financier to mold small companies into a system, and (2) a major technological innovation not controlled by the company. Western Union's optimal strategy was to allow the entry of specialized companies so long as they could be prevented from uniting to form a significant system. Companies would tend to come in to meet specialized needs or to exploit new inventions. Since they would have a small market share, a price war against them would be more costly for Western Union than for the small company. However, an attempt to buy out each small competitor would provide a strong incentive for companies to enter in order to benefit from the Western Union purchase. The best strategy was to allow the small companies to coexist so long as they remained small and localized, but to take action if they began expanding. Some of the small companies were captive companies of the railroads and thus could not be driven out by price wars. The small companies provided a nucleus for rapid expansion if a well-financed company should attempt to construct a serious competitor to Western Union. There was some incentive to merge the small companies to gain a share of the monopoly profits or to force Western Union to pay a high price for the new company, but exceptionally good financing was necessary for an effective threat. The new company would need the ability to withstand extensive construction costs and a long period of price warfare in order to be a credible challenger to Western Union.

The first serious competitive challenge to Western Union occurred in 1874. The Atlantic and Pacific Telegraph Company was one of the many small competitors that had slowly expanded. It had completed lines from New York to Buffalo in 1867 and had continued building west. It reached Omaha, Nebraska, in 1874 and began arranging connecting links with other small companies, including a key Ogden–San

Francisco line owned by the Central Pacific Railroad. The company was controlled by Jay Gould. Gould's financial strength made it possible for the new company to seriously challenge Western Union. Western Union at the time was controlled by the Vanderbilt family. The competition between Western Union and the Atlantic and Pacific was part of a larger struggle for control of major companies between Gould and the Vanderbilts. The new company hired a top operating executive from Western Union, began an active program of mergers (including the purchase of the Franklin Telegraph Company) to establish nationwide connections, undertook a considerable amount of new construction, and cut prices by 40 percent. It appears that Gould's primary goal was to drive down the price of Western Union stock and gain control of the industry via Western Union rather than to make money on his telegraph company. Western Union did not cut below Atlantic and Pacific's prices and the new company gradually took business away from Western Union, doubling its revenue between 1875 and 1876. In 1877 Atlantic and Pacific initiated another sharp rate reduction, which brought a merger offer from Western Union. Western Union first purchased a small percentage of Atlantic and Pacific stock and formed a cartel with the new company, then arranged a complete merger in 1881.[22]

The limited capacity of the small competitive lines and the time required to construct new lines, as well as the problem of finding routes for new lines, protected Western Union from rapid loss of its business even with a well-financed, determined competitor such as the Atlantic and Pacific Company. Western Union's strategy of continuing with high prices until the entrant proved its determination to be a serious competitor, then offering favorable merger terms, appears to have been profit maximizing. By only offering merger to a large company, Western Union limited the incentive for new companies to enter in order to be bought out. By purchasing the entrant while it was still a small percentage of Western Union's size, Western Union avoided the danger of losing control of the industry. By refusing to engage in a price war with the entrant, Western Union avoided the significant losses to its profitable business which could have been necessary to inflict relatively minor financial losses on the much smaller competitor. If the competitor had been less well financed, it might have been optimal to engage in a price war with it, but in this case the entrant had financial resources similar in magnitude to those of the dominant firm.[23]

The second potential threat was the displacement of the telegraph as a whole or Western Union's method of operating the telegraph by a major technological change. Minor inventions were not a threat because the inventors would be likely to sell them to Western Union rather than to attempt to use them to establish a competitive system. Western Union's route and systems barriers to entry protected it from serious technological challenge from all but truly major inventions. Thus it was necessary for Western Union to monitor the progress of technology and attempt to purchase inventions that appeared to have promise for its business.

The most promising direction for telegraph improvement was the prospect of carrying more than one message on a telegraph wire at one time. Because much of the investment of a telegraph system consisted of wires and poles, a duplex or multiplex system would provide a tremendous cost savings to Western Union or pose an extreme threat to Western Union's market power if it were controlled by a competitor. To guard against technological displacement, Western Union set up a research group in 1869 to undertake research on its own and to evaluate inventions done outside the company for applicability to telegraph operations. A practical duplex system was invented in 1868 by J. B. Stearns. Western Union purchased his patent for a reported $250,000 and brought it into regular use by 1872. In 1874 a team of Thomas Edison and George Prescott developed a quadruplex system (four simultaneous messages) which Western Union installed on 13,000 miles of wire by 1878.[24]

Another inventor offered Western Union a patent on a method of carrying information over wires in 1877. Western Union considered the asking price of $100,000 too high and refused the offer.[25] The invention came from a teacher of the deaf named Alexander Graham Bell.

Evaluation of Telegraph Development

The telegraph industry development indicates a failure to maximize industry profits because of overconfidence in the degree of protection provided by the Morse patent. This caused the patent holders to underinvest in barriers to entry and led to considerable competition by 1850. From 1850 on, the industry was concerned with renewing monopoly control through mergers and cartel agreements. Mergers took place both between directly competing companies to eliminate price competition, and between geographical monopolies to produce a unified system. Western Union developed a protected place in the industry through favorable railroad contracts and an active merger program. It was then propelled to industry dominance by the Civil War. Western Union was close to a textbook monopoly after the Civil War, with primary competitive protection coming from its control of favorable routes and its extensive system, but its pricing policy induced the entry of small firms which provided a nucleus for a competitive system.

The telegraph history indicates that it is difficult for a cartel to maximize long-run industry profits. The differences in expectations among cartel members, as well as the opportunities for member firms to advance their own interests by disregarding the interests of the cartel, make it difficult for the cartel to invest in barriers to entry. The telegraph industry was in effect a cartel from the beginning because of the organizational structure of the patent partnership. Thus the individuals concerned found it difficult to agree on the proper actions to be taken, leading to a short-run strategy with little concern for investing in market protection. Once Western Union gained control, a long-run profit-maximizing strategy was begun.

4 The Telephone in the United States

The Telephone Challenge, 1876–1879

The rapid growth in telegraph demand and resulting heavy investment in new telegraph wires focused the attention of inventors inside and outside of Western Union on the economic value of an invention that would allow more than one message at a time to be transmitted on a telegraph wire. The first practical duplex method (two simultaneous messages) was developed by a non–Western Union inventor, J. B. Stearns, in 1868. His patent was purchased by Western Union for a reported $250,000 and brought into regular use by 1872. In 1874 a team of Thomas Edison and George Prescott developed a quadruplex apparatus (four simultaneous messages), which was installed on 13,000 miles of wire by 1878.[1]

Elisha Gray, the electrician for the Western Electric Company (at that time the manufacturing arm and partial subsidiary of Western Union), spent several years during the early 1870s developing a sophisticated system for multiple transmission based on the ability of a wire to carry various musical pitches at the same time. He expected eventually to reach fifteen simultaneous circuits on the wire. After having been granted patents on various aspects of his "harmonic telegraph," Gray decided that the apparatus he was developing could be used to transmit speech as well as musical tones. Accordingly, he filed a caveat for a patent (preliminary to actual application) on a device "to enable persons at a distance to converse with each other through a telegraphic circuit just as they now do in each other's presence or through a speaking tube" on

February 17, 1876.[2] In the caveat he described his proposed apparatus but had not yet transmitted speech through it.

While Gray was developing his "harmonic telegraph," a teacher of speech to the deaf named Alexander Graham Bell was also experimenting with a multiplex telegraph. Bell had made extensive studies of sound patterns in his efforts to teach the deaf to speak. He thought he could build a multiple telegraph substantially better than the Stearns version but lacked financing and electrical expertise to carry out his experiments. The fathers of two of his students, Thomas Sanders and Gardiner Hubbard, offered to finance his experiments in return for sharing any proceeds equally among the three of them. Hubbard was an attorney with considerable knowledge of the telegraph industry and its opportunities. Although the original commercial incentive was to develop a multiplex telegraph, Bell himself was more interested in a telephone from the beginning.

After experiments that convinced Bell that he had the right ideas, but without actually transmitting speech, Bell filed a patent application February 17, 1876, the same day as the Gray caveat but slightly earlier. The patent was granted March 3, 1876, and Bell succeeded in transmitting speech on March 10. However, the device used by Bell on March 10 was quite different from the one described in his February 17 application and was similar to the device described in the February 17 Gray caveat. This fact has led to extensive controversy by historians of the period over the question of whether Bell's working device was really his own work or appropriated from Gray. It has been established that the patent examiner conveyed some information from the caveat to Bell (described as illegal but common practice) but it is uncertain how much information was transmitted or what use was made of it by Bell.[3]

Bell's 1876 patent was entitled "Improvement in Telegraphy" and only vaguely claimed the invention of the telephone. The telephone-related claim stated: "The method of, and apparatus for, transmitting vocal or other sounds telegraphically, as herein described, by causing electrical undulations, similar in form to the vibrations of the air accompanying the said vocal or other sounds, substantially as set forth."[4] In later litigation the 1876 patent was held to be the controlling telephone patent. Bell continued to work on the telephone aspect of his invention, and after successful experiments transmitting actual speech, he filed a new patent application January 15, 1877 (granted January 30, 1877) in which he described his previous patent as relating to "a method of and apparatus for producing musical tones by the action of undulatory currents of electricity, whereby a number of telegraphic signals can be sent simultaneously over the same circuit, in either or in both directions, and a single battery be used for the whole circuit" and clearly laid claim to the telephone, asserting that his new invention was "the electrical transmission . . . of articulate speech and sound of every kind, whether musical or not."[5]

Bell and his financial supporters believed that his invention was significant and valuable. However, the direct commercialization of it presented a number of obstacles. The most important of the obstacles was Western Union, with its financial strength, dominance of the telecommunications industry, and patents on communications technology including Gray's directly competitive work. Bell's telephone was superior to the telegraph in its ability to provide direct voice communication without a skilled operator, but it was limited in distance to about twenty miles while telegraph communication extended across continents.

Without further developments, the telephone was complementary to the telegraph rather than competitive, because it could not be used for long-distance communication but could be used to gather local messages for transmission by telegraph. Local telegraph service had been largely unprofitable except for the Gold and Stock Telegraph, a subsidiary of Western Union that transmitted price quotations from the stock exchange to broker's offices. Although the telephone would not displace a significant amount of telegraph business at first, the complementary nature of telephone and telegraph made Western Union a potential telephone power. If the inventions of both Bell and Gray were held to be valid and independent, Bell would find Western Union a formidable competitor.

Bell's position was weaker than Morse's initial position because of the existence of an established company with patent rights to related devices. The experience with the Bain telegraph patent had indicated that a patent can have a significant competitive effect even if it is later held to be invalid. The existence of a potentially valid patent claim prevents the courts from issuing an immediate injunction and the actual trial process takes so long that the competitive battle may be over before the litigation is finished. In the telegraph case, O'Reilly had been able to regain control of his lines and build a large number of new lines under the Bain patent before it was finally declared invalid. In the telephone case, Western Union had the potential to overpower a commercial effort by

Bell while the patent situation was being litigated, even if Bell should finally be upheld.

Given the competitive threat posed by Western Union and its patents, as well as the ordinary business problems of raising capital and marketing a new product with uncertain demand, the low-risk strategy for Bell was to offer to sell his patents to Western Union for a suitably high price. As with Morse's earlier attempt to sell the telegraph patent to the government, the inventor placed a higher valuation on his patents than did the entity to whom they were offered. The actual reasons for Western Union's refusal to purchase the Bell patent are uncertain. In the past the company had been willing to pay a high price to eliminate competition through generous merger offers. However, this case was somewhat different in that Bell was only offering a patent and not an existing competitive company. Many prior inventions related to the telegraph had turned out to be insignificant and Western Union would have found it unprofitable to purchase every invention offered to it to avoid potential competition. The fact that the telephone as it existed then was not directly competitive to the telegraph business, combined with Western Union's own patent claims on telephone technology, reduced the company's enthusiasm for paying $100,000 to Bell. Another potential factor in the decision was that considerable animosity had arisen between Gray and Bell over the credit each of them deserved for the invention.[6] If Western Union depended on Gray for information regarding the relative significance of the work done by Bell and Gray on the telephone, the company would have been likely to gain an optimistic picture of its standing. Western Union refused Bell's offer to sell the patents for $100,000 and, rather than negotiate for a lower price, Bell and his supporters decided to commercialize the invention on their own.

The first attempt at commercialization was undertaken as a manufacturing company rather than as an operating company. Instead of setting up a telephone exchange, the company offered to lease telephone instruments to customers who would be responsible for stringing their own wires to those with whom they wished to communicate. Each line would have a phone on each end and would provide direct private-line service between the two points without the ability to switch calls among lines. In May 1877 the Bell group advertised sets of two telephones for lease at a rate of $20 per year, if used for social purposes, and $40 per year, if used for business purposes. The Bell group would be responsible for keeping the instruments in repair and the subscriber would be responsible for erecting his own wires, which were estimated to cost from $100 to $150 per mile. Additional phones for the same line would be provided at $10 per year each. The primary advantages claimed for the telephone over the telegraph were that no skilled operator was required and that communication was more rapid. The phones received relatively rapid acceptance and, by the end of July 1877, 778 phones had been leased. At that time the Bell Telephone Company was formed and Hubbard was designated trustee with power to act for the others.[7]

Among the best customers for the early Bell telephones were the New York brokers who had been customers of Western Union's Gold and Stock Telegraph Company. To protect its small but profitable subsidiary, as well as to expand its local communications business, Western Union formed the American Speaking Telephone Company in December 1877, as a subsidiary of the Gold and Stock Telegraph. The new company was given rights to the patents of Elisha Gray and Amos Dolbear. In addition, Western Union hired Thomas Edison to develop telephone improvements. He soon invented the carbon transmitter, which provided voice quality superior to that provided by either Bell or Gray and gave Western Union a technical advantage in the competition.[8]

Spurred on by the competition and the desire to occupy major cities and gain the competitive advantages of being first, both companies pushed telephone development hard during 1878. It became apparent very quickly that maximum benefit could be derived from the new invention by connecting each local phone to a switching center so that all subscribers could converse with each other, rather than by the initial plan of private wires between phones. However, setting up exchanges required extensive investment in wires and switching equipment, straining the underfinanced Bell Telephone Company. Rather than raising the needed capital for telephone exchanges directly, Bell began to franchise local operating companies as had the Morse patentees thirty years earlier. The local company would rent telephones from Bell and pay a percentage of its stock for the privilege of operating under the Bell patents. The first operating company formed was the New England Telephone Company (organized February 1878), which paid 50 percent of its stock to Bell Telephone. Immediately after its formation, New England Telephone opened the first telephone exchange, in Hartford, Connecticut. American Speaking Telephone used its financial power and widespread connections to establish exchanges around the country. During the first half of 1878, telephone exchanges were established by

American or Bell or both in San Francisco, St. Louis, Chicago, and other major cities.[9]

In mid-1878, Bell Telephone was reorganized to bring in additional capital and professional management. Two-thirds of the stock of the new company was issued to the patent holders and the remaining one-third was sold to Boston capitalists for $50,000. The shares sold for cash had greater voting rights than the shares issued to the patentees, allowing control of the company to be exercised by the capitalists. The sale of one-third of the new company for $50,000 gave an implicit value of $100,000 for the prior company in mid-1878. Thus even after a very successful year, which had included the establishment of operating companies and the installation of 10,000 telephones, the patents plus other company assets were only valued at $100,000, the price refused by Western Union for the patents alone in 1877. About the same time as the financial reorganization, Theodore Vail was induced to leave his position as superintendent of the Railway Mail Service and become general manager of Bell Telephone.[10] The company structure and management policies established by Vail have remained influential to the present time.

The reorganization and appointment of Vail gave ultimate control of the company to capitalists unconnected with the invention as such, and operating control to a very capable professional manager. Bell himself took no part in the company's business affairs. As a result, the company was able to develop a systematic plan for the profitable development of the telephone business without the complications caused by disagreements among the patentees in the loosely organized Morse patent association.

Under Vail's leadership, Bell Telephone pursued its fight with Western Union during late 1878 with three tactics: (1) continued rapid development of telephone exchanges, (2) a patent infringement suit filed in September, and (3) technical development for improved quality. Although Bell pushed its telephone coverage rapidly, Western Union was installing phones at a greater rate during 1878. Bell's ability to install phones was hindered by Western Union's control of the telegraph lines. Western Union refused service to places that installed Bell telephones and thus effectively prohibited Bell installations in hotels, railways, and newspaper offices that depended on Western Union service. The patent infringement suit did not produce an immediate injunction against Western Union's use of the telephone and was expected to develop into a long legal battle because of the complexity of the competing claims. The technical improvement program was quite successful. The most notable result was the acquisition of rights to the Blake transmitter, which eliminated the Bell phones' competitive disadvantage to the Edison transmitter.[11]

Both companies continued to install telephones at a rapid rate during 1879, while negotiations were begun to settle the patent suit out of court. After extensive negotiations, a settlement between the two companies was reached in November 1879. Western Union agreed to withdraw from telephone service for seventeen years, to sell its network of 56,000 phones in fifty-five cities to the Bell Company, to transfer its telephone-related patent rights to Bell, and to pay 20 percent of the cost of any new Bell telephone patents for seventeen years. In return Bell agreed to stay out of the telegraph business, to turn over to Western Union all telegraph messages that came within its control unless directed otherwise by the customer, and to pay Western Union 20 percent of its rental on telephones. Bell paid Western Union approximately seven million dollars over the life of the agreement.[12]

Western Union's bargaining position was weakened by a competitive attack on its telegraph business during the telephone competition. Jay Gould, a leader of the Atlantic and Pacific Telegraph which had settled with Western Union in 1877, organized the American Union Telegraph in May 1879 at the height of the competition between Western Union and Bell.[13] Gould also began buying control of the Bell licensees. The Gould efforts increased the risk to Western Union of pursuing the competition with Bell, because if Gould should gain control of Bell companies and win the patent suit, he could create a combined telephone-telegraph company that would present a formidable challenge to Western Union. However, a settlement agreement that tied Bell to Western Union would deprive Gould of potential business for his telegraph system and strengthen Western Union against the immediate telegraph threat.

Much discussion has taken place over the significance of the agreement and over which company better recognized the future importance of the telephone industry. That investors saw the agreement as extremely favorable to Bell can be inferred from the price of Bell stock during the negotiations. In March 1879, after the patent suit had been filed but while competition was still vigorous, Bell stock sold for around $50 per share; in September a rumor of settlement drove it up to $300;

the day after the agreement was completed it jumped to $1,000 per share.[14] However, given that the Bell patents were valid, the settlement appears favorable to Western Union. As financial historian J. Warren Stehman put it:

> The American Speaking Telephone Company had no controlling patents, nothing but claims based upon Elisha Gray's invention (which was not being used in the development of a practicable telephone) and patents upon details. Nevertheless, the Bell interests bought its physical property at a fair valuation and paid large annual sums of money for minor inventions which, admitting the validity of the Bell basic patents, no one other than the Bell people could have used. It seems to be a case in which the financially powerful Western Union overawed a young and weaker company.[15]

Further insight is given to the negotiations by the personal account of Theodore Vail, who described the settlement as "the most important single event in the history of the telephone business." As Vail described the negotiations:

> The negotiations hung on the condition denying to the Bell interests the right to connect their exchanges by means of toll lines. Few had faith in the future of the toll lines or their value as compared with the private lines, but if long distance conversation should be developed the Western Union feared it might be a menace to the telegraph business.
>
> The conferees of the Bell were divided about the toll business; some of them tired of the contest, preferred half a loaf in peace and comfort, rather than a struggle for a whole loaf; if yielding would bring about a settlement some were willing to yield. To me the idea of yielding the toll line use meant the curtailment of our future—the absolute interdiction of anything like a "system."[16]

Both companies had strong incentives to reach agreement. The rapid consumer acceptance of telephones during the short period of competition indicated that local telephone service would be a large and potentially profitable market. The elimination of competition between the two companies would allow increased total profits in that market. The competing patent claims of Western Union and Bell could lead to invalidation or narrow construction of both sets of patents if the litigation were pursued. However, the combination of both sets of patents would be far stronger against outside challengers. Thus a merger of the telephone patents could eliminate the immediate competition between the two companies and also reduce the risk of additional entrants.

At the time, the telephone and telegraph were complementary. The telephone could not be used for long-distance service because of technical limitations, and the telegraph was infeasible for local service because of the need for a skilled operator. The telephone could increase the value of telegraph service by substituting for messengers to collect and deliver messages. The telegraph could increase the value of telephone service by allowing long-distance communication. The complementary nature of the products also made independent telephone and telegraph companies a threat to each other. A telephone company could enter the telegraph business and use its control of the telephones to route business to its own telegraph lines. A telegraph company could enter the telephone business and use its control of the telegraph lines to generate business for its own telephones. Either case would require the other company to enter the complementary product in order to protect itself.

An additional complication was the likelihood that long-distance telephone service would become feasible. That would make telephones directly competitive with the telegraph. If telephone service were more expensive than telegraph service, the telephone would be a specialized competitor that would attract customers who placed a high value on voice contact and immediate response. If the long-distance telephone should become inexpensive relative to telegraph, it would take away the bulk of the telegraph business, leaving only customers who placed a high value on a written record of the communication.

Disregarding patent considerations, Western Union was clearly in a stronger competitive position than Bell. Its established position in telegraph service gave its telephones (connected with the telegraph) an advantage over Bell telephones (only in places not served by telegraph.) Western Union's management ability and financing also allowed it to act more quickly than the new company. Given Western Union's competitive use of the telegraph system and the infeasibility of long-distance telephone, Bell needed to establish its own telegraph network in order to reach competitive equality with Western Union. The difficulty of establishing a telegraph network and the possibility of being driven out of business by Western Union with continued competition made it advantageous for Bell to file the patent suit against Western Union despite the risk it posed to Bell's patents.

From Western Union's point of view, the patent litigation had only two likely outcomes, both of them unfavorable: (1) the courts might uphold Bell completely, enjoin Western Union from further telephone operations, and reduce the value of its existing telephone operation, including patents, to nearly zero; (2) the courts could invalidate the Bell patents and leave the telephone business open to entry by any who desired it. If the telephone company combined with a rival telegraph, the first outcome would threaten Western Union's main business as well as eliminate its telephone business. The second outcome would leave Western Union in a strong position to profit from the telephone business by carefully using its financial power, offices, franchises and rights of way, and connections with its telegraph system to promote its telephone business, but the industry would be likely to encounter substantial competition and to fail to earn monopoly level profits. A third possibility was the invalidation of the Bell patents and the upholding of the Western Union patents to such an extent that Western Union could bar entry, but that would have been considered extremely unlikely. However, the combination of the patents of the two companies would have been considered very likely to produce an effective barrier to new entry.

From Bell's point of view, pursuit of the litigation to a clear victory would save the 20 percent royalty to Western Union, and would allow the company to seek rebates from Western Union for business delivered to the telegraph or to enter telegraph operations directly. However, even if Bell believed its patents would be upheld, it could justify the royalty to Western Union as an investment in barriers to entry. The company gained Western Union's explicit promise not to enter the telephone industry for seventeen years, Western Union's acknowledgment of the validity of the Bell patent, and access to telephone improvements developed by Western Union. The explicit promise not to enter was valuable because it eliminated Western Union's incentive to develop noninfringing apparatus and reenter even if the Bell patents were upheld. The acknowledgment of the validity of Bell patents was in Western Union's interest because of its royalty on Bell rentals, and was introduced as evidence in later Bell patent litigation. A ruling that invalidated the patents or construed them so narrowly that entry could occur without infringement would have been far worse for Bell than the settlement. Bell had none of the established business advantages of Western Union, and except for having a slight head start in establishing exchanges, it could expect to earn only normal profits if free entry were allowed.

The conditions in the industry made agreement possible even with different expectations on the outcome of the patent litigation. The advantages to Bell from settlement were enough to make the settlement in Bell's interests even if the patent should be upheld and broadly construed. A cooperative telephone monopoly paying a 20 percent royalty to Western Union was potentially better for Western Union than open competition if the Bell patent should be invalidated.

Given the infeasibility of long-distance telephone service, the settlement established a duopoly in telecommunications with the market divided according to local service (Bell telephone) and long-distance service (Western Union telegraph). Both companies were dependent on the other. Both gave an explicit promise not to enter the other's market. Telegraph barriers to entry were strengthened by Bell's agreement to deliver telegraph messages to Western Union. Telephone barriers to entry were strengthened by Western Union's acknowledgment of the validity of the Bell patent and sale to Bell of the Western Union patents. The agreement protected the interests of both parties. However, if Bell should be able to develop long-distance service, it would free itself of dependence on Western Union and become the dominant company. Thus as Vail recounted, the most difficult part of the negotiations concerned Bell's rights to develop its own long-distance telephone network. Western Union's acquiescence to Bell's demand that it have the right to build long-distance lines provided Bell with a powerful incentive to develop feasible long-distance communications and free itself from dependence on Western Union.

Bell Patent Monopoly, 1879–1894

The Western Union agreement eliminated Bell's strongest competitor and provided additional defense against the entry of other competitors. It left Bell close to the position of a textbook pure monopolist until 1894. However, Bell could not be certain at the time of the agreement of how broadly the telephone patents would be construed or if they would be upheld at all. It therefore had an incentive to create additional barriers to entry even during the patent period. In addition, the company could expect competition at the expiration of the Bell patents and the Western Union agreement. If barriers to entry could be created rapidly for the same cost as more gradual creation, little attention to barriers to entry would be expected until the end of the patent period, followed by substantial barrier investment at the end of the patent period. However, barriers generally take time to create, and thus it was optimal for the company to be planning its defenses against post-patent competition all during the patent period.

The opportunities for creating barriers to entry included: (1) obtaining additional patents to cover all details of telephone equipment, (2) developing long-distance technology and lines, (3) occupying all desirable locations with telephone exchanges, and (4) vertically integrating manufacturing and telephone operation.

A large number of patents would provide back-up protection for the main Bell patents. If the main patents were invalidated, the telephone could still be protected through patents on various design features. Improvement patents provided an opportunity to extend the life of the monopoly by preventing competitors from using recent technology.

Long-distance lines would be important to create a telephone system, to tie together the operating companies, to minimize dependence on Western Union, and to increase the value of the telephone network. If the system consisted of a group of local operating companies connected only by Western Union lines, then it would be susceptible to Western Union entry after the agreement expired. The individual companies would also have little advantage over a competitor with only a local exchange. However, if the companies were tied together with a long-distance network, then a Bell subscriber gained access not only to other local subscribers but to all subscribers in the Bell system. This would provide a competitive advantage over an isolated competitor in the same way that the Western Union system had a competitive advantage over isolated telegraph companies.

Extensive building of the telephone system would prevent entry by denying customers to the new company. Few customers would be likely to leave the established network to join a new exchange until that exchange was large enough to provide access to a large number of subscribers. The easiest way for a competitor to begin would be to build where there was no Bell exchange or to supply customers who were priced out of the market by the Bell system. The wider the Bell coverage, the more difficult it would be for a competitor to enter the market. However, wide coverage also meant lower prices to induce more customers to join the network, which could make this a very expensive barrier to create.

An additional opportunity to create barriers to entry was to directly control manufacturing rather than purchasing from manufacturing companies. Vertical integration can create a barrier to entry by requiring a potential entrant to master all stages of production rather than only one.[17] Of the four potential barriers to entry, Bell put primary emphasis on patents, put less emphasis on long-distance lines and vertical integration, and largely ignored extensive development as a barrier to entry.

During the monopoly period, prices could be set at the profit-maximizing level without regard for the entry incentives created. If the patents and other barriers proved to be inadequate protection, prices could be reduced, but it would be very costly to reduce prices in advance to decrease the incentive to enter. This policy could be expected to cause rapid growth of the telephone system in the initial years as the company installed telephones in new areas, with lower growth in later years as the market approached an equilibrium between the monopoly price and the quantity demanded at that price. If prices were reduced either through Bell action or competition, that action could be expected to induce another round of rapid growth in the total number of telephones.

Following the 1879 Western Union agreement, the Bell company was reorganized as the American Bell Company in 1880. The reorganization included a six for one stock split and the sale of additional stock to bring the company's total capitalization up to five million dollars. American Bell then established permanent license agreements with its associated companies to replace the previous temporary ones. A license was granted to operate a telephone exchange in a particular area in return for 30 to 50 percent of the stock in the company formed. Under the terms of the license, local capitalists would be responsible for raising the money necessary to establish the exchange. Telephone instruments would be rented to the local company by American Bell and title to the instruments would remain in the parent company. The local company was not allowed to build long-distance lines outside of its territory and could only connect with other exchanges through the parent company or its designated representative. In addition, the companies agreed not to participate in any telephone business not licensed by American Bell. This provision prohibited a local company from developing its own instruments.[18] The license contracts developed a close tie between American Bell and the operating companies while maintaining the ability of the operating companies to tap the local and regional capital markets of the time. Although a national capital market was beginning to emerge, it would have been difficult for American Bell to raise the extensive amounts of capital necessary for a nationwide telephone system.

American Bell publicly took the position that its monopoly control was thoroughly established in 1880. The annual report that year stated:

> With a thorough occupation of the principal cities and towns by our licensees, the ownership of the broad patents covering the use of the speaking telephone, and the control of nearly all of the inventions for apparatus necessary to the telephone business which have yet been made, the danger of competition with our business from newcomers seems small.[19]

However, the company also recognized the need to develop additional protection and devised a three-part plan to create barriers to entry. Vail later recounted the program as follows:

> The Bell Company, from the commencement of the business, intended to control the business. The intent is not only claimed by all who were parties to the management at the time, but it is shown in every record of every transaction in the course of the business. *One system, one policy, universal service* is branded on the business in the most distinct terms.
>
> . . . The policy of the Bell Company, as it was originally thought out, embraced in its broad lines all the features that exist today—Control of Patents . . .
>
> Control of the individual exchanges, through direct ownership, partnership, or short contracts—and particularly through the rights reserved to the Bell Company to connect these exchanges with each other and control all intercommunication either with each other or with the outside world.
>
> The control of the Manufacturing Department, through license and after through ownership, was essential to standardization and uniformity of instrument . . .
>
> A study of the evidence will show that the Bell Policy was to establish the business on the same lines as if it was all done direct by the Company with its own funds, only substituting a corporation with its Manager, . . . performing the duties of a District Manager.[20]

The first aspect of Vail's program to "control the business" was to extend the company's patent protection. At the time of the Western Union settlement, Bell had twenty-four patents of its own and forty-two patents and applications from Western Union. Bell's patent portfolio still left it vulnerable to significant improvements in telephone apparatus that could be used by a competitor. As Vail stated the concern:

> The telephone patent never did cover the whole telephone business . . .
>
> In the early telephone history there was a great open field for suggestive invention and it was soon taken advantage of. The greatest danger that threatened this new industry was control by unassociated people of the necessary devices for working.[21]

In 1881 Bell established an Engineering Department to conduct research and experimentation and to evaluate outside inventions for relevance to the telephone. The Engineering Department concentrated its own resources on applied research, but was quick to purchase rights to more fundamental inventions. As Bell's patent protection strengthened, so did its bargaining power with inventors of telephone-related apparatus because of the difficulty of commercializing the invention outside of the Bell system. All Bell employees were required to turn over rights to telephone-related inventions to the company. Bell's patent attorney of the period described the program as follows:

> It has been the practice of the company, watching the history of the art, . . . to select everything in the way of conductors, apparatus, or telephones that it has seemed would be of benefit to the system at large, and either to buy them outright, which was nearly always done in the first instance, or else to buy sufficient rights under them to enable the American Co. and its system to carry on the telephone business with high efficiency everywhere.[22]

The patent program produced a total of nine hundred patents for Bell by the expiration of the original patents in 1894.

The high profits of the Bell system provided strong incentives to enter the market and many companies attempted to establish competitive exchanges. Bell was vigilant in its defense of its monopoly and filed some six hundred patent infringement suits during the life of the original patent. The combination of Bell and Western Union patents created a strong enough presumption of infringement that in most cases Bell was able to get an injunction to stop the business of the competitors. While most potential competitors left the field after receiving an injunction, some pursued their claims through full trials and appeals. Defenses brought by potential competitors against the infringement charges included: (1) operation with a new invention not covered by the Bell patents, (2) invalidity of the original Bell patent because it did not describe an instrument capable of transmitting speech, and (3) invalidity of the original Bell patent because of prior work of other inventors.

The various Bell patent cases were consolidated for a joint Supreme Court decision. In March 1888 the Supreme Court ruled that the original Bell patent was valid and should be given a broad interpretation, covering telephone communication in general rather than one particular method of accomplishing telephone communication. The decision came on a four to three vote. The dissenters held the view that the original patent was invalid because of evidence that a noncommercialized version of the telephone developed by Daniel Drawbaugh had preceded Bell's invention.[23] Because the strongest challenge to the original Bell patent was eliminated with the Western Union agreement, the close decision suggests that patent protection might have broken down in the industry if the Bell-Western Union suit had been pursued to completion.

The second aspect of Vail's program to develop industry control was to develop and control long-distance lines. The first telephones were used with telegraph wires and were limited to a range of twenty miles. Telegraph wires were single wires with the ground used as a return circuit. The limited range of early telephones made many franchisees relatively unconcerned about rights to long-distance lines, but Vail saw the long-distance lines as an important method of control over both the local franchisees and potential competitors. He stated:

> The first big effort was in multiplication of intensive local development and the conserving of the future, the preservation of all the future possibilities in a way to make them most effective when evolved. The telephone man of the day wanted the exchange rights. So he was given the exchange rights, and all other rights were reserved. I wish you particularly to note this in regard to interconnecting lines. In part payment for the exchange rights, . . . the right was reserved to run extra-territorial lines or toll lines into and connect them with any exchange.[24]

Short interexchange lines, beginning with a Boston–Lowell link, were established early in the monopoly period. The company then discovered that a two-wire circuit (without using the ground as a return) greatly reduced the interference on the line and increased the range of effective communication. In 1885 American Telephone and Telegraph Company was formed as a wholly owned subsidiary of American Bell to develop long-distance service throughout the country. That year the first interstate commercial line was completed to provide service between New York and Philadelphia. An experimental New York–Boston line was completed in 1884 but commercial service did not begin until 1889. Considerable effort was expended to improve long-distance technology and to build long-distance lines during the later years of the patent monopoly. By 1893 telephone communication was possible between New York and Chicago and between Augusta, Maine, and Washington, D.C.[25] Telegraph service remained the primary method of long-distance communication, but the existence of a long-distance telephone network provided an alternative to the telegraph which was growing in significance at the end of the monopoly period. The long-distance lines created a semblance of a national telephone system superior to isolated local exchanges even though most subscribers made few if any long-distance calls.

The third step in creating barriers to entry was to integrate equipment manufacturing and telephone service. Independent manufacturing of the equipment needed by Bell would tend to encourage independent invention and potentially lead to loss of control. Vail stated:

> In the manufacturing, there were a large number of independent individuals and firms, each developing apparatus to be used in connection with the development of the telephone.
>
> If this continued, the owners of the telephone invention and the business would be controlled by the outsiders and the manufacturers and those who were employed to develop it.[26]

The ownership of a manufacturing source would protect inventions and provide an additional barrier to entry. If the original policy of licensing small independent manufacturers to produce under the Bell patents had continued, those products would have been potentially available on an equal basis to all when the patents expired. If Bell owned the supply side, a new company would have to overcome the combined entry barriers of manufacturing and telephone exchange service.

Rather than developing its own manufacturing organization, Bell purchased an interest in Western Electric, the supply arm for Western Union. Western Electric had experience manufacturing both telephone and telegraph equipment, and had excess capacity after it lost the orders of the Western Union telephone company. Western Electric also owned many valuable patents related to telephone and telegraph apparatus. In July 1881 American Bell purchased 40 percent of the stock of Western Electric. The company was subsequently reorganized and an additional 20 percent of the stock of the reorganized company was issued to American Bell in exchange for a permanent, exclusive license to manufacture telephones under the Bell patents. The reorganization gave Bell control of the company. Bell later increased its ownership to 100 percent. In an 1882 contract between the two companies, American Bell agreed to use only Western Electric to manufacture telephones, and Western agreed to only sell telephones manufactured under either Bell or Western Electric patents to Bell-licensed companies. The contract remained in force until 1908 when it was amended to allow Western Electric to sell phones to non-Bell licensees.[27] The agreement completed the patent pooling begun by the Western Union agreement and eliminated interference between Bell and Western Electric patents.

The three-part program developed a protected position for Bell at the end of the patent monopoly. The first line of defense was the array of nine hundred patents that provided opportunities for extensive legal challenges to any potential entrant. The integrated system of manufacturing company, local operating companies, and long-distance lines, each pledged to deal exclusively with other Bell companies, presented a further obstacle to any potential entrant in an industry in which the value of the service depended on the number of people with whom connections were available.

The primary system revenue during the period came from charges for local exchange service, with a much smaller amount from long-distance toll calls. The local charges were set by the local operating company and varied from place to place. The rates were determined by a simple price discrimination scheme based on fixed annual charges, with higher rates for businesses than for residences. The operating companies rented the instruments from American Bell at a relatively constant price over the period. The average rent per telephone charged by the parent company ranged from a low of $10.76/year per phone in 1886 to a high of $11.68/year in 1885. At the operating company level, prices were gradually increased over the monopoly period. Average annual operating revenue rose from $70/year per phone in 1885 to $90/year per phone in 1894. While the rental paid to American Bell remained fairly constant, other operating expenses were rising at about the same rate as the prices. Net earnings per telephone thus remained fairly constant between $30 and $35 per year during the monopoly period.[28]

The number of phones increased rapidly for the first few years and then moved slowly once the initial equilibrium was reached. Although various sources differ substantially as to the actual number of phones installed in the years before 1885, there is agreement that the percentage growth was large in each of those years and that approximately 142,000 phones were installed by 1885. The 1885 figure appears to be approximately an equilibrium number of telephones for the price charged and the quality of the service. For the next nine years, the number of phones grew at an average rate of 6.3 percent per year, reaching a total of 270,000 phones in 1894.[29]

Development was concentrated in the centers of the cities, with little effort given to rural lines or small towns. The centers of major cities were clearly the most lucrative, but it does appear from hindsight that the Bell system missed profitable opportunities in the less densely populated areas. It is possible that the structure of the system caused the actual retail price to be above the profit maximizing price. This would occur if the rental charged by the parent company was set at a monopoly level and then used as an input price for the operating company to again compute a monopoly-level final price. The failure to service small towns and rural areas created a reservoir of unsatisfied demand, which provided a strong inducement for the entry of new firms once the patent protection was weakened. It is likely that if the Bell system had used a more sophisticated pricing system to serve the outlying areas at lower gross margins than the central cities, it could have both increased current earnings and better protected itself against entry. However, demand at the time was uncertain and the high current profitability and protected position of the Bell system provided little incentive for managers to experiment with alternative pricing schemes.

The monopoly was highly profitable. During the seventeen-year monopoly period, a total of $16 million was invested in the business from the sale of stock and the conversion of bonds into stock. During that time, cash dividends totaled $26 million. Stockholder equity at the end of the period was $38 million. The average annual return on investment was approximately 46 percent. The period was one of general deflation with (nominal) interest rates below 5 percent.[30]

Soon after the Western Union–Bell agreement, Western Union merged with Gould's American Union Telegraph and reestablished its monopoly of the telegraph service. Western Union continued its pre-telephone program of expansion and reduction in prices as costs were reduced through technological progress. For most of the telephone monopoly period, Western Union was relatively free of competition from the telephone. The expanding long-distance telephone network provided some competition toward the end of the period, but the high charges and limited distance of the telephone left most long-distance messages to the telegraph. Western Union revenues rose from $9.9 million in 1878 to $25.0 million in 1893, for an annual growth rate of 6.2 percent compounded over the period. Western Union's percentage growth in revenues was higher in the early part of the period than in the later part. During the monopoly period, the Bell system developed from a tiny company to an equal of Western Union. Bell's revenue was insignificant in 1878, 54 percent of the Western Union revenue in 1885, 68 percent of the Western Union revenue in 1890, and 98 percent of the Western Union revenue in 1895.[31]

Beginning in the 1850s efforts were made to lay a transatlantic cable to provide direct telegraph communication between Europe and the United States. After several failures, a successful cable was completed in 1866.[32] The cable was operated under a cartel of United States and foreign companies in which Western Union was the United States agent for the receipt and distribution of international messages. The editor of the *New York Herald,* an extremely heavy user of international telegraph service, was instrumental in organizing the Commercial Cable Company to lay a competitive cable in order to bring down prices. Commercial Cable completed its transatlantic cable in 1884. After a price war in which the international telegraph rate dropped from 50¢/word to 12¢/word, a new cartel was formed including Commercial Cable and the price was set at 25¢/word, low enough to remove the incentive of the newspapers to sponsor another cable but apparently still profitable. To expand service beyond New York without becoming dependent on Western Union, Commercial Cable purchased a small bankrupt telegraph company known as Postal Telegraph and used it to merge together several small companies and build new lines. It never attained complete coverage of the country nor seriously challenged Western

Union's domestic business, but it opened offices in major cities and built enough lines to provide a constraint on Western Union's pricing freedom. In 1904 John Moody listed Commercial Cable/Postal rather than Bell as the most important rival of Western Union.[33]

The growth of the Bell system into a profitable large company and the entry of Commercial Cable into the telegraph business greatly reduced Western Union's market power by the end of the Bell–Western Union agreement. Distant communication could take place over the telephone long-distance lines or Postal Telegraph lines as well as through Western Union. While Western Union was still a strong and profitable company, its own market segment was declining relative to the telephone and it could not easily enter the telephone business. An attempt by Western Union to end its cooperation with Bell and establish its own telephone company would be likely to induce a partnership between Bell and Postal. Such a partnership could increase the difficulty of entering the telephone market and also seriously challenge Western Union's telegraph business. Western Union was also suffering managerial difficulties at the end of its agreement with Bell that reduced its ability to restore its place in the industry. Consequently Western Union settled for maintaining its profitable telegraph business rather than risking entry into the telephone technology. Even with hindsight, it is difficult to say whether the discounted value of profits for Western Union would have been higher with a move into telephone, but the decision relegated Western Union to a secondary position in the telecommunications industry.

The Decline of Monopoly Power, 1894–1907

The two fundamental Bell patents expired in 1893 and 1894. Potential entrants still faced the obstacle of nine hundred Bell patents covering every aspect of the telephone and related equipment. The patents were of uncertain significance because they had not been litigated. So long as the fundamental patents prohibited entry, there was no occasion to test the detail patents. The significance of a patent is in general uncertain until some company infringes it and litigation is begun. After litigation, the patent may be invalidated or construed narrowly. On the other hand, the patent may be upheld and construed broadly with a resulting loss to the entrant. Thus the existence of unlitigated patents increases the risk to potential entrants, but does not necessarily bar entry. The significance of unlitigated patents as a barrier to entry increases with the risk aversion of potential entrants and with the losses that would be incurred in a forced exit, as well as with increases in the potential entrant's estimate of the probability of the patent being upheld.

Besides patents, potential entrants faced the barriers of the lack of experienced companies to manufacture telephone equipment, and the existence of Bell operating companies in all the major cities with many of them connected through Bell long-distance lines. Bell's vertical integration meant that the potential entrants could not use the same source of supply as the Bell operating companies, but there were many companies with electrical expertise and the willingness to respond to a profitable opportunity to develop equipment for telephone entrants. The occupation of the major cities prevented potential entrants from beginning in the most lucrative markets, since Bell would not allow interconnection between its telephones and competitors'. Interconnection was technically feasible, but so long as the Bell system was much larger than its competitors, Bell gained a competitive advantage by refusing interconnection.

While it would be practically impossible for a new entrant to establish a system equal to Bell's in a short period of time, the systems advantage to Bell was reduced by the fact that most telephone subscribers communicated with a relatively small number of people. Although the value of having a telephone would rise with the total number of people connected, the amount of increase would depend on the existing subscribers' desire to communicate with the new subscribers. If a new entrant could connect to a small but homogeneous subgroup of the population, its service would be valuable despite the limited total number of phones in the system. If the Bell system and the new competitor generally served different social classes in the same city (as often happened during the era of competition), the advantage of having the two systems interconnected could be relatively small. The fact that telegraph service was far more pervasive than long-distance telephone service at the expiration of the patents also reduced the systems advantage by allowing subscribers to an isolated telephone exchange to conduct long-distance business via telegraph.

Overall, the barriers facing potential entrants in 1894 were high but not insurmountable. Incentives to enter were strong because of the high profitability of the Bell system. The failure of Bell to provide service to the outlying areas of the major cities and the smaller towns created an opportunity for new entrants to begin business without having to induce existing Bell customers to switch to a new system. Bell apparently underestimated the significance of the expiration of the fundamental patents, and did not reduce its prices in any significant way until after entry actually began. The quantum drop in barriers to entry should have induced a price reduction to decrease the entry incentives. Uncertainty over the degree of patent protection Bell possessed in 1894 and over the rate of entry at various prices made it very difficult to determine the optimal price.

Three kinds of entry into the telephone business occurred. The first, and by far the most important, was the formation of new companies to establish commercial telephone systems in the hope of making a profit. The second type of entry was through the establishment of a cooperative to develop a mutual telephone system. The mutual systems were organized into exchanges with central switchboards, as were the commercial systems, but were owned by their subscribers. The third type of entry was through the establishment of farmer lines. Farmer systems were very informally organized telephone lines to connect the farmers of an area. These consisted of a single line in which all conversations reached all subscribers or multiple lines with a simple switchboard operated by one of the farm families. Farmer lines were organized for as little as $100 in capital.[34] In some cases, the telephone signal was even carried over existing barbed-wire fences. Although the farmer lines were unimportant competitively and had a very small percentage of the total phones installed, their existence was an important indicator of the strength of telephone demand and the feasibility of unsubsidized telephone service in sparsely populated areas.

In 1894 eighty commercial systems and seven mutual systems were established, primarily in areas without Bell service. By the end of that year, the new entrants had a total of 15,000 phones installed (an average of 172 phones per system) for a combined market share of 5 percent. According to H. B. MacMeal, a historian of the independent telephone movement, the first commercial exchange was established in Noblesville, Indiana. The initial impetus for the competitive exchange was the local banker's anger over the Bell company's arrogance and poor service, rather than a rational expected profit calculation. After extensive searching, the banker and other local businessmen found a company willing to risk a Bell suit to produce telephones for them. They used a telegraph switchboard to avoid infringing Bell's patents for telephone switchboards. The new company earned a 30 percent profit on its $10,-000 investment the first year after cutting the Bell rates in half. The profitable Noblesville example encouraged other entrepreneurs to form non-Bell telephone companies.[35]

The success of the first Bell challengers induced widespread entry in the years after the basic patents expired. Competitive systems were established at an increasing rate between 1895 and 1900. A total of 199 new commercial telephone systems were begun in 1895, 207 in 1896, 254 in 1897, 334 in 1898, 380 in 1899, and 508 in 1900. The systems extended to all areas but were concentrated in the Midwest. By 1902 three thousand non-Bell commercial telephone systems had been established. Five states (Illinois, Indiana, Iowa, Missouri, and Ohio) had over two hundred telephone systems each. The first companies acquired telephones and equipment from a wide variety of electrical manufacturers, but the rising demand for telephone equipment and the declining probability of being shut down by patent infringement suits caused several companies to develop specialized telephone equipment manufacturing capability. The most important entrants into the manufacturing side of telephone competition were Kellogg Switchboard and Supply (formed 1897), Automatic Electric Company (formed 1901), and Stromberg-Carlson Telephone Manufacturing Company (formed 1902).[36] These companies developed technical expertise in telephone technology and reduced or eliminated the competitive disadvantage of the new entrants from their inability to purchase from Western Electric.

As the competitive movement gained strength, the Bell challengers moved from the small towns to the major cities. City competition brought about the first form of telephone regulation. The cities granted charters to the competitors that contained fee payments to the cities, limitations on rates to be charged, and/or restrictions on mergers with other companies. For example, the Indianapolis charter to the New Telephone Company required a payment to the city of $6000 per year plus $2 for each telephone above 6000 and limited the rates that could be charged to $40/year for businesses and $24/year for residences. At that time the Bell company was charging $72 for business and $48 for residences. In Toledo, the entrant's franchise called for forfeiture of the franchise if the company should assign or sell its property to any telephone system in the city of Toledo. The antimerger restriction in Toledo and other places prevented the entrants from securing a city franchise in order to capitalize its nuisance value in a merger with the Bell system.

Companies organized to compete in New York and in Boston failed to secure the right to lay cables in the subways and thus could not establish a competitive system in those cities.[37] The need for a city franchise put others at a competitive disadvantage to the Bell system. The franchise barrier to entry became more significant over time as city governments became more reluctant to grant franchises to additional competitors.

Direct competition with Bell in the major cities required far more capital than was needed by the initial small town entrants. In many cases capital was easy to raise because of the expected high profitability of the enterprises. In an era of few securities regulations, this led to fraudulent or misleadingly promoted schemes that made profits for the promoters but left financially weak companies to face the Bell competition. A common practice was for the promoters of the new telephone company to own the construction company that was given the contract. An inflated price would then be charged for construction and paid out of the capital raised, giving the promoters substantial profits regardless of the success of the company. In many cases the nominal capitalization was far greater than the funds received from investors, which apparently induced unsophisticated investors to believe the company possessed more physical capital than it did. In some cases the companies issued bonds for the valued of cash received and issued stock to the bondholders as a bonus without actual payment. The fixed interest nature of the bonds then left the companies with no flexibility to meet even normal contingencies of business, much less a sustained rate war.[38] Questionable financial practices reduced the ability of the independents to provide long-term rate reductions and tarnished the public image of the independents. Although the long-term effect was to weaken the independents as viable competitors, the short-term effect was to increase the level of competition, because the promotional profits from organizing new telephone companies made entry profitable regardless of the long-term outlook for the companies.

Because of the high Bell prices and limited service areas, the first competitors were successful with isolated systems. As the competitive movement broadened, the disadvantages of isolated exchanges became more apparent. The new entrants began considering ways of adding long-distance lines to connect independent exchanges in various cities. The first independent long-distance lines were small-scale undertakings to connect adjoining independent exchanges. The first large-scale effort to establish an independent long-distance network came with the formation of the Telephone, Telegraph, and Cable Company in 1899. The company was backed by substantial capitalists of the day and announced plans to buy control of independent telephone companies and build a competitive long-distance network. The company quickly bought control of several small independent telephone systems and the Erie Telegraph and Telephone Company. The Erie Company had started just before the expiration of the Bell patents and had been sued for patent infringement. The companies reached an out-of-court settlement in which Erie took over Bell patent rights in widely scattered areas (Texas, Ohio, Minnesota) in exchange for 22 to 30 percent of the stock in the operating companies. This arrangement made some of the Erie operating companies Bell licensees without complete Bell control. Less than a year after its organization the Cable Company claimed control of 115,000 phones, about 11 percent of the total phones in the United States at that time.[39]

By 1900 telephone competition was widespread. The independents controlled 38 percent of the phones installed in the United States. Although the independent phones were spread among some two thousand tiny companies, they provided direct competition to almost all Bell operating companies and the potential for a merger into a unified nationwide competitor. The rapid progress of the well-financed Erie Company appeared to pose a threat to Bell's status as the only large-scale interconnected system. However, the Bell system was aware of the danger to its market power and profits, and responded with a program to maintain its position in the industry. Bell's response took four forms: (1) patent infringement suits, (2) reduced prices, (3) increased development of the long-distance network, and (4) mergers with key competitors.

The first hope of the Bell system was that the patent monopoly could be extended, either through a broadly interpreted controlling patent or through the combination of many less significant patents on details and improvements. The Berliner microphone patent was Bell's best candidate for a controlling patent. The Berliner patent had been applied for in 1877 but it had not been granted until 1891 and therefore was valid until 1908. The Bell system claimed that the patent covered microphone action in general rather than Berliner's specific design for a microphone. If the courts would uphold such an interpretation, all telephones would infringe the patent and the Bell monopoly could be extended until 1900. However, Bell was also worried about the ability of the Berliner patent to withstand challenge as shown by the following letter to the president of Bell in 1891:

> The Bell Company has had a monopoly more profitable and more controlling—and more generally hated—than any ever given by any patent. The attempt to prolong it 16 years more by the Berliner patent will bring a great strain on that patent and a great pressure on the courts. This has nothing to do with the validity of the patent, or the duty of the Courts to sustain it . . . Patents which would stand ordinary litigation have been known to give way under great strain, if they turn on questions when it is humanly possible to take an adverse view.[40]

If the Berliner patent should fail to provided adequate protection, the second hope was to extend the patent monopoly by enforcing the large number of minor patents. Such a policy could prevent the manufacture of some types of equipment and increase the expense and risk of manufacturing telephone equipment in competition with the Bell system. The Bell patent attorney wrote in the 1892 annual report:

> Referring to patents generally, I may report that the people of the West, so far as I could see, seem to be rather frightened at the idea of infringing patents, and the great number of the projected and organized rival concerns alter their construction in order that they may not infringe the patents of the Western Electric Company; but the outcome of this is that they thus sell and use inferior instruments.
>
> From what I observed in Chicago, it appears to me that the policy of bringing suit for infringement on apparatus patents is an excellent one because it keeps the concerns which attempt opposition in a nervous and excited condition since they never know where the next attack may be made, and since it keeps them all the time changing their machines and causes them ultimately, in order that they may not be sued, to adopt inefficient forms of apparatus.[41]

When new companies began actual competition with Bell, Bell filed patent infringement suits against the manufacturers of competitive equipment. Bell filed five suits in 1894 and twenty-three more in 1895. The suits did manage to keep the "opposition in a nervous and excited condition," but were unable to stop the competitive movement. The Berliner patent was upheld but construed narrowly to cover only Berliner's specific method of making a microphone. As that method had been superseded, the patent had little effect on competition. Another important setback for Bell was a court ruling that the important switch-hook patent (which allowed a call to end by hanging up the phone rather than turning a switch) was invalid because it possessed "no novelty or invention in any one of the eight claims presented."[42] Many other patents were either invalidated or narrowly construed by the courts. The Bell patents were generally interpreted to prevent direct copying of Bell equipment, but not to prevent the manufacture of functionally similar equipment with a different design. Thus the competitive manufacturers were generally able to design around the Bell patents and continue in business. The patent portfolio increased the cost and the risk to an entrant, but did not in itself bar entry.

The second response was to cut prices. A price cut reduces the incentive to enter and provides a signal to a potential entrant that retaliatory action may be taken if it enters. When entry occurs, the established company must choose between cutting prices to drive out the entrant and allocating the entrant a share of the market. If a large number of potential entrants exist, accepting the entrant into a cartel is not an effective strategy because it will only induce further entry. An extremely large number of potential telephone entrants existed. Thus, if there had been no chance of patent protection, we could have expected to see a drastic price cut immediately after the expiration of the fundamental patents. Two complications arose in computing the optimal amount of a price cut. The first was the chance of further patent protection. The higher the probability of significant patent protection, the less we would expect to see the price cut. The second problem came from the structure of the Bell system as a holding company with incompletely controlled operating subsidiaries. The parent company did not directly control the pricing policy of the operating companies and thus could not impose a uniform response to competition on its subsidiaries.

The organizational structure was significant because of differences in the incentive to cut prices for the parent company and the operating companies. For the parent company, a successful entrant in one city would be likely to induce entry into other cities. Thus it would be desirable to cut prices and attempt to drive out the entrant or at least make it relatively unprofitable. However, from the operating company's point of view, once a competitor has entered its market, other potential competitors are likely to seek other cities for entry rather than to put a third telephone company in one city. Thus there is an incentive to accept the new entrant and attempt to reach a cooperative agreement, at least until all similar cities have at least two telephone companies.

The Bell parent company began reducing its prices to the operating companies immediately on the expiration of the patent. The rental charge per telephone to the operating companies was almost constant for the last nine years of the monopoly. It averaged $11.12 per year per phone during that time. It was then reduced to $7.78 in 1894, $4.36 in 1895, $3.74 in 1896, $3.46 in 1897, and $2.90 in 1898. The operating companies also gave price reductions to their customers, but by a far smaller percentage amount. While the rental to operating companies was cut by 74 percent between 1893 and 1898, the total operating revenue per telephone of the entire system dropped 28 percent (from $90/year to $65/year) over the same period.[43]

Most operating companies cut prices after competitors entered, but retained prices above those charged by the competitive company. Thus the Bell company did not begin a price war and signaled its preference for higher prices to the new company. In a few cases, the Bell company started an all-out price war with drastic price cuts below cost, but those appear to have been rare. Although some individual operating companies encountered financial difficulties because of local price wars, the Bell system as a whole earned normal competitive profits or higher during the competition. Annual dividends of the parent company continued during the competitive era, but at a slightly reduced rate (from $18 per share during the monopoly to $15 per share during competition). Return on investment for the system as a whole declined from 46 percent during the monopoly to 8 percent during the years 1900–1906.[44] The price cuts undertaken by Bell were responses to actual competition rather than defensive moves to reduce the incentives to enter. By cutting prices to close to the level of actual competitors, the operating companies prevented the total loss of their business, but they did not take away the incentive for further entry. During the early competitive period, Bell was selling market share by continuing to hold prices above the competitive level.

The third response by the Bell system was to increase its continuing efforts to connect the operating companies with long-distance lines. The greater the significance of long-distance lines to customers, the greater the advantage the Bell system would have over isolated competitors. A letter to the president of American Telephone and Telegraph (then the Bell parent company as a result of the purchase in 1900 of American Bell with AT&T stock to switch the corporation from Massachusetts law to New York law) in 1901 stated:

> I take it that it is extremely important that we should control the whole toll-line system of intercommunication throughout the country. This system is destined, in my opinion, to be very much more important in the future than it has been in the past. Such lines may be regarded as the nerves of our whole system. We need not fear the opposition in a single place, provided we control the means of communication with other places.[45]

Effective long-distance communication was hampered by the problems of creating a usable telephone repeater or amplifier. A simple mechanical repeater had been developed early for the telegraph because it only needed to distinguish between two signals and regenerate the appropriate signal with accuracy. Consequently telegraph communication had been limited only by the length of the wires constructed. However, with the telephone, the voice was transmitted in an infinite set of gradations and a far more complex amplifier was needed. Despite considerable Bell efforts, a practical mechanical repeater was not available until 1906, and repeaters did not come into widespread use until the vacuum tube allowed the construction of instruments far superior to the mechanical repeater. A vacuum-tube repeater system was used to establish the first transcontinental telephone line in 1915.

Lacking an effective repeater, the early Bell system efforts were directed toward reducing the resistance of the wire, primarily by increasing its thickness. Aided by many minor improvements, the Bell system completed operational lines between the East Coast and St. Louis, Minneapolis, and Kansas City by 1898.[46] Soon afterward, communication was established with Denver by using a special line with extremely thick wire. In 1900 an important new method of increasing the communication distance without repeaters became available through the invention of the loading coil by Michael Pupin of Columbia University. Pupin's patent rights were quickly purchased by the Bell system. The Pupin loading coil extended the feasible range of communication and also reduced the construction cost by allowing the use of thinner wire.

The Pupin patent allowed Bell to extend its long-distance network and put potential competitors at a technological disadvantage to Bell. By gradually increasing the extensiveness and quality of its long-distance network during the competitive period, Bell was able to reduce the ability of isolated competitors to maintain effective competition.

The fourth Bell response to competition was a program of selective mergers. No general merger program was undertaken during the early competitive period, because favorable merger terms to new competitors would only induce greater entry. Merger restrictions in some of the competitive city franchises and strong antipathy toward merger or cooperation with Bell from many of the independents also prevented a program of widespread mergers. Two classes of companies provided potentially attractive merger opportunities for Bell. The first group consisted of actual or potential long-distance companies. The large number of isolated competitive companies provided an opportunity for a well-financed company to merge them together and build competitive long-distance lines to establish a significant competitive system. If the Bell system could buy out potential long-distance entrants, it could gain some protection from competition at far less expense than by purchasing individual operating companies. The second class of attractive merger candidates was the group of equipment manufacturers. Although barriers to entry in equipment manufacturing were not extremely high, each company had developed specialized expertise and the independents would be put at a disadvantage if they were unable to continue purchases from their accustomed supplier. In addition, the suppliers took primary responsibility for defending users of their equipment against Bell patent infringement charges. A carefully arranged acquisition in which the new subsidiary was not granted patent rights could benefit Bell in its legal attempts to prove that the competitors had infringed Bell patents.

The first major merger target of the Bell system was the Telephone, Telegraph, and Cable Company. Soon after the Cable Company's acquisition of the Erie Telegraph and Telephone Company, two of its major financial backers had withdrawn their support, allegedly at the request of J. P. Morgan, who hoped to gain control of AT&T and did not want another integrated system.[47] With the loss of key financial supporters, the Cable Company was in need of additional financing to proceed with its ambitious plans. In 1901 Bell offered to secretly finance the purchase of control of the Cable Company through C. W. Morse, president of the American Ice Company and a director of the Cable Company. The Boston News Bureau report of September 3, 1901, stated:

> Mr. Morse believes that the time is ripe for telephone competition in the East, and if nobody else is willing to undertake it, he is willing to go ahead, provided the terms of his offer are accepted . . . He believes there is room for two companies . . .
>
> There is suspicion in some quarters that Mr. Morse's offer is in behalf of Bell interests, but we think this suspicion is unfounded.[48]

The public should perhaps have had some reason to doubt Morse's commitment to competition as he had just finished monopolizing the New York ice supply and drastically raising prices. However, Morse's offer was accepted and the stock was then turned over to the Bell companies. The minority stockholdings of the Cable Company were also acquired and the company was dissolved. The purchase of the Cable Company did not immediately give Bell direct control over Erie because the Cable Company had placed some of its Erie shares in a trust agreement to secure loans to Erie. However, AT&T liquidated the Erie debt and reorganized the Erie companies as regular Bell operating companies.[49] Because of the financial needs of the Cable Company before its acquisition, the price paid was not high and therefore the purchase did not induce additional entry. It is likely that Bell acquired control more cheaply by using a secret agent than if it had made a direct merger offer because the stockholders would be willing to sell control of the company in order to bring in additional financing for expansion at a lower price than to allow the dissolution of the company. The purchase of the Cable Company together with the acquisition of the Pupin patent rights at about the same time stopped interest in a competitive long distance company.

Two merger attempts with independent manufacturers were less successful. In 1902 Milo Kellogg, the founder of Kellogg Switchboard and Supply, became seriously ill and turned the management of the company over to Wallace DeWolf. Kellogg also gave DeWolf power of attorney for his controlling interest in the company. DeWolf sold the controlling interest to agents of the Bell system under an agreement in which ownership of the stock was to be kept secret and DeWolf was to remain head of the company. Bell controlled the company secretly for eighteen months while Kellogg continued to bid on projects to supply independents. Meanwhile Kellogg was engaged in patent litigation with Bell and provided the guarantee against infringement to all companies purchasing its equipment. The alleged reason for the secret control was to present a sham defense of the patent claims, allowing Kellogg to lose them, and then to shut down the operating companies that were using the Kellogg apparatus. At least one Bell-Kellogg trial was held during the period of secret ownership, resulting in an order to modify the Kellogg equipment to avoid infringement. MacMeal states:

> When Judge Taylor learned a year afterwards, that at the time when the Bell counsel was pleading for judgement against the Kinloch and Kellogg companies, the Kellogg company had already been under Bell control for several months, his amazement was unbounded.[50]

Milo Kellogg regained his health and attempted to recover his stock without success. A suit was then instituted by the minority stockholders against Bell and the sale of Kellogg's stock was set aside on antitrust grounds.[51] Kellogg remained an independent manufacturer.

A second merger attempt also failed on antitrust grounds. The Bell system attempted to purchase control of Stromberg-Carlson in 1907 but the sale was rescinded after antitrust action was brought by the New York state attorney general.[52] No further attempts to gain control of the independent manufacturers were made. With the decreasing importance of patents and antitrust prohibitions on acquiring the independent manufacturers, Bell abandoned the attempt to cut off supplies to the competitive telephone companies. In 1908 the Bell-Western Electric contract was amended to allow Western Electric to sell telephone equipment to companies which were not licensed by AT&T.

The four Bell system responses (patents, price cuts, long distance lines, and mergers) slowed its loss of market share but did not stop it. New competitors continued to enter and existing competitors continued to expand throughout the period 1894–1907. The proportion of telephones controlled by the independents rose from 19 percent in 1897 to 44 percent in 1902, and then slowly increased to 49 percent in 1907.[53] However, the market share figures indicate an overly pessimistic assessment of Bell's market power in 1907. By 1907 entry was becoming more difficult and long-distance lines were becoming more important. There were no significant undeveloped areas of the country and the major cities already had two telephone companies in most cases. Many of the independent telephones were in small towns and rural areas, which had less opportunity for profit than the major cities. Independents had been relatively unsuccessful in the largest cities of the East Coast. The independents were in general very small and had limited capital resources. Bell's patent position in long-distance transmission made it unlikely that the independents could establish a fully competitive integrated system. The initial investor enthusiasm for establishing competitive telephone companies had worn thin after years of modest and in some cases nonexistent profits. The decreasing strength of the competitive movement, despite its increasing market share, opened up the opportunity for profitable large-scale mergers with independent operating companies to reestablish control of the industry.

The price reductions, selling efforts, and service improvements of the competitive era created a dramatic surge in telephone demand. Bell's annual average growth rate of 6.3 percent during the last ten years of the patent monopoly jumped to 21.5 percent during the first ten years of competition, while the entire market was growing even more rapidly. The total number of telephones doubled during the last ten years of monopoly but was multiplied by a factor of twelve during the first ten years of competition. After seventeen years of monopoly (1894), the United States had a limited telephone system of 270,000 phones concentrated in the centers of the cities, with service generally unavailable in the outlying areas. After thirteen years of competition, the United States had an extensive system of six million telephones, almost evenly divided between Bell and the independents, with service available practically anywhere in the country.[54] The rapid growth of Bell and the independents during the competitive period suggests that Bell probably underestimated the elasticity of demand for telephones and maintained a higher than optimal price during the monopoly period. A lower price under the monopoly could well have expanded the market enough to create higher immediate profits as well as reducing the incentive for competitors to enter after the expiration of the patents.

Evaluation of Telephone Development

The early telephone story provides insight into the creation and the destruction of market power. Western Union's monopoly was broken,

Bell's monopoly was established, and Bell's monopoly was weakened during the period. In retrospect, it is clear that Western Union made a costly mistake in turning down Bell's original telephone patent. However, at the time the telephone patent would have appeared to be a marginal investment for Western Union. Telephone prospects and the significance of the Bell patent for telephone development were both uncertain. Telephones were not a direct substitute for telegraph but a complementary product. They replaced messengers or personal travel, not telegraph lines. The telephone would displace potential telegraph demand from local service if the operation of telegraph could be simplified sufficiently to make local service feasible. The telephone would displace telegraph lines if its range could be extended. Thus the complementary nature of the products could be transformed into competition between them by progress in either technology.

Western Union's displacement by Bell as the dominant telecommunications supplier indicates the difficulty of protecting monopoly power against technological innovation. Western Union was a progressive firm for its time; it undertook research on its own and purchased outside technology relevant to telegraph operations. It was willing to pay a high price for useful inventions, such as the Stearns duplex telegraph. Yet it could not feasibly purchase every patent related in any way to telegraph. It lost its monopoly by refusing to purchase a patent on a device on which its own inventors were working and that could not be directly used in the telegraph network or provide direct competition to the telegraph. The inherent uncertainty of the significance of new inventions and new patents prevents any firm from correctly estimating the value to be placed on all new developments. Consequently, even if all parties offer to sell their inventions to the dominant firm first, an abundance of technological opportunities will result in some major inventions being passed over and used to challenge existing market power.

The telephone development also illustrates the ability of a firm to use temporary monopoly power to generate long-range barriers to entry. The telephone instrument itself has few natural barriers to entry. There are no significant economies of scale in telephone set manufacturing, and many companies had the technological ability to produce the sets. The original patent provided a broad and controlling monopoly on the manufacture of telephone sets for its duration. It would have been difficult to maintain monopoly power in telephone sets after the expiration of the patent if the sets had simply been manufactured and sold as ordinary electrical appliances. The temporary patent monopoly on telephone sets allowed Bell to establish a monopoly of telephone exchange service in which much greater natural barriers to entry existed. Right of way to Bell for telephone wires was freely granted because Bell was the only company that could provide telephone service during the monopoly period. The city franchises and systems advantages from its customer network then provided continuing market power after the patent expiration made telephone sets freely available. The initial control of the telephone sets and resulting control of telephone exchanges allowed Bell to also control long-distance communications through restrictions on the rights of Bell licensed exchanges to provide long-distance service through any means other than the Bell network. The interrelated system of telephone sets, local exchanges, and long-distance service all pledged to connect only with other Bell companies, provided continuing market power after the patent expiration.

A far more significant and enduring monopoly was created by using the patent to erect a telephone system than would have occurred by simply selling telephone instruments at a monopoly price. If telephone instruments were sold at a monopoly price to all comers during the patent period, it is likely that competitive exchanges would have been developed during that period and competitive equipment supply afterward, restricting the effects of the original Bell patent to its lifetime. Bell's careful use of its patent period to develop a defensible position in telephone service has allowed it to benefit from the original patent for over a hundred years.

The reduction in monopoly power after the expiration of the patent resulted from overconfidence in the significance of the follow on patents and inadequate coverage of the country with telephones during the monopoly period. Bell only offered service to the centers of the major cities. The early competitors provided telephones to small towns and rural areas that had been unable to secure service from Bell rather than providing direct competition to Bell's exchanges in the major cities. The success of their initial efforts caused them to move into direct competition. It appears that the failure to offer more widespread service was simply a mistake in estimating the potential profits to be made in sparsely settled areas. It would have been in Bell's interest to offer more widespread coverage even at cost in order to prevent easy initial entry after the expiration of the patents. Bell's failure to serve the smaller towns or to reduce prices sufficiently after the expiration of the basic patent allowed extensive entry. However, Bell's interconnected system, long-distance network, and dominance in the largest cities still gave it substantial monopoly power in 1907. Although it had only about half of the total phones, it remained the dominant firm with significant advantages over its many tiny competitors.

5 The Telephone and Telegraph in Europe

Telephone and telegraph development occurred with much greater government involvement in Europe than it did in the United States. In most countries, private companies played an initial role but were eventually absorbed into a government monopoly. The competition between government firms or licensing agencies and private firms provides insight into the distinctions between private and public market power. The impact of government control of telegraph on the development of the telephone illuminates the process of major technological change under government monopoly. The rate of telephone development in European countries provides perspective on the significance of the change in United States market structure from monopoly to competition for the United States telephone development rate.

This chapter is necessarily selective in its coverage of the varying conditions in different European countries. Events have been chosen that provide perspective on the United States development. No attempt is made to provide a complete account of European telecommunications. Considerable attention is given to events in Great Britain and more summary accounts of the French, German, and Swedish experience are provided.

The Telegraph in Great Britain

The initial British telegraph industry was based on the apparatus of Edward Cooke and Charles Wheatstone. The Cooke-Wheatstone telegraph patent application was filed in December 1837, four months prior to the time the Morse application was filed in the United States. The Cooke-Wheatstone telegraph was based on signals from the electric current causing a needle to point to a particular letter. The coding and decoding of letters into dots and dashes by the operator, required in the Morse apparatus, was unnecessary in the Cooke-Wheatstone device. However, the Cooke-Wheatstone system was much more expensive to install than the Morse system. It required five wires at first and two wires in its later versions, while the Morse device required only one wire. The Cooke-Wheatstone patent was upheld against other British inventions of the same time period. Morse's attempt to secure a British patent was fought by Cooke and Wheatstone and the patent was denied,[1] giving Cooke and Wheatstone a patent monopoly in the British market.

At the time of the Cooke and Wheatstone invention, the British railway system was just beginning. An initial high value use for the telegraph was to signal the passing of trains and help prevent accidents on single-track railroads. Cooke and Wheatstone made an agreement with the Great Western Railroad to build a thirteen-mile line in May 1838. The line was completed in July 1839 (five years prior to the experimental Baltimore–Washington line in the United States) and worked satisfactorily to notify the railroad of passing trains. However, the railroad decided that other means were adequate to its signaling needs and refused to extend the telegraph line. Cooke and Wheatstone then turned to a contract similar to the later Western Union contracts with the United States railroads. The telegraph company agreed to provide free service to the railroad in exchange for use of the right of way. The inventors signed similar contracts with a number of railroads and began telegraph development in earnest.

By 1845, 550 miles of railroad line were completed or under construction, a slow development rate compared to the United States. However, the slowly growing system of railroad telegraphs was profitable and the extensive railroad construction activity suggested that future telegraph profits could be substantial. Consequently, Cooke and Wheatstone were able to sell a 72 percent interest in their patent rights to outside capitalists for £115,000 in December 1845. Following the sale of the patent rights, the Electric Telegraph Company was formally incorporated to develop the telegraph in Britain.

Development of the British telegraph system remained slow. Prices were set at a high level. The resulting volume of commercial business was low and profits were below expectations. The first dividend was not paid until June 1849. After providing the government with emergency communications service during 1848, Electric Telegraph unsuccessfully sought a government subsidy as a firm that was providing a socially necessary but unprofitable service. By 1850 the company had wires along 2,215 miles of railroad, but 7,231 miles of railroad were open and 4,795 more miles were under construction. Thus the exclusive contracts with the railroads failed to provide competitive protection because of the large unserved areas. In the United States, in contrast, rapid development under competition in the late 1840s had produced 12,000 miles of wire by 1850 (over five times the British total) despite a later start.[2]

The development of an alternative to the Cooke-Wheatstone patented device brought another company into the telegraph business in 1850. Under competition, the maximum rates within England fell from 10s. to 4s. Rather than amounting to an unprofitable price war, this reduction so stimulated demand that total industry profits rose. Both companies undertook substantial construction and extended the range and value of telegraphic communication. Demand was further increased by the successful laying of ocean cables, beginning with an England–France link in 1851 and followed soon afterward by an England–Holland link and an England–Ireland link. Between 1851 and 1855 Electric Telegraph tripled its wire mileage while its volume of business increased by a factor of seven.

The profitability of the companies provided incentive for new entry. In addition, the failure of the duopoly to test the demand at lower price ranges left open the possibility that the 4s. price charged by the duopoly was above the profit maximizing level just as the 10s. price charged by the monopoly had been. Because the duopoly had occupied most of the railroad routes with exclusive contracts, a new entrant needed right of way on the public roads as well as noninfringing telegraph apparatus. In 1860 the United Kingdom Telegraph Company was activated to establish additional telegraph service. The UKTC paid £12,000 in stock for rights to the Hughes telegraph and announced its intention to establish a uniform 1s. rate for a message of twenty words anywhere in the United Kingdom. The existing cartel attempted to block entry by defeating or delaying the necessary parliamentary authorization for the new company to lay wires along public roads. Electric Telegraph registered £6,000 in legal and lobbying expenses opposing the right of way, while the UKTC registered £11,709 in expenses to gain passage of the bill.[3]

The UKTC secured construction authorization in 1862 and completed a line from London to Manchester the next year. Additional cities were added soon afterward. The duopoly dropped rates to 1s. for routes served by the UKTC and maintained higher rates for other routes. The UKTC did not find the uniform 1s. rate as profitable as it had expected. In 1865 a new cartel was formed, including the UKTC, with distance-based rates of 1s. for under 100 miles and 2s. for between 200 and 300 miles. Even after the new cartel was operative, the new rates were much lower than before the entry of the UKTC. Profits and dividends increased after the formation of the new cartel. Electric Telegraph raised its dividend from 7 to 10 percent on nominal capitalization between 1863 and 1865. The 10 percent rate was the highest dividend paid by the company. During the cartel period, profits on real physical investment for the three companies were approximately 15 percent. Electric Telegraph remained the dominant firm in 1868 with 58 percent of the domestic messages of the three companies, while the Magnetic carried 28 percent and the UKTC carried 14 percent.[4]

The actions of the three companies were consistent with expected cartel behavior. The major efforts against the UKTC were directed at preventing it from establishing its system through lobbying in Parliament and strict enforcement of right-of-way advantages. These efforts increased the costs of the UKTC. They also increased its risk because of the possibility of failure to secure the right of way. After the UKTC began business, the established companies cut the price on the competitive routes to equal the competitor's price, preventing the company from taking all the business on those routes. However, after the lines were built, it would have been difficult to drive the UKTC completely out of business because it would be rational for the company to remain in business so long as it could cover the variable costs without regard for the fixed investment in its lines. The willingness of the companies to form a cartel in 1865 increased profits without causing very much inducement for additional entry because of the established patent position of the three companies and four years of low profit experience of the UKTC.

The development of British telegraph structure up to 1868 was similar to the development of industry in the United States. The initial monopoly based on patents was destroyed by the invention of noninfringing apparatus and the entry of new competitors. Each new competitor brought a time of reduced rates and then peaceful coexistence. By 1865 the industry had settled into a stable cartel, as had the American industry, but the British companies did not actually merge together as the American companies did in 1866. Both the intensity of competition and the level of cartel profitability were lower in Britain than in the United States. The invention was of lower value in Britain than in the United States because of the relative compactness of the country. The greatest value of the American telegraph came from the savings of days of travel time for such long and important routes as New York–New Orleans and New York–San Francisco, especially when the telegraph preceded the railroad. Britain had much shorter distances and earlier coverage by railroads, making the total time savings in communicating with domestic points less important. The British companies also incurred higher costs of construction from the lack of timber for poles and more extensive development of the land, which made rights of way more significant. These factors combined to make the profit maximizing price much closer to the level of average costs in Britain than in the United States and thus reduced the incentive to enter the industry or break down a cartel agreement.

Prior to the formation of the cartel in 1865, there had been some agitation for government ownership of the telegraphs. Support for nationalization came from newspaper owners who believed they could get lower rates and better service under government ownership, from the British Post Office, and from the telegraph companies. The formation of the cartel in 1865 and the resulting rate increase revived public interest in government operation of the telegraph. The supporters of nationalization claimed that the systems could be purchased for approximately £2.5 million, the outstanding capital assets of the companies, and that with a uniform 1s. rate the service could pay operating expenses and interest on the purchase debt. They predicted substantial cost savings and service improvements from Post Office control of the telegraph. After extensive debate, a rather vague bill was passed in 1868 authorizing the government to take over the telegraphs and make them a part of the Post Office but not appropriating any money for the action and leaving the price to be paid unspecified.

The government proposed to pay the highest price that had been quoted on the stock exchanges for the stock of each of the companies, but the companies refused the offer and requested payment of twenty-five times annual profits instead. The government finally agreed to pay twenty times the 1868 profits of the companies. The price paid was estimated to be 34 percent for real property and 66 percent for goodwill. The Post Office sought and received an explicit monopoly authorization for telegraph operations. The generous arrangements with the companies, together with previously ignored costs such as purchasing rights from railroads and international cable companies, raised the total price paid for the telegraph system from the originally estimated £2.5 million to £10 million. The companies were allowed to capitalize their highest year of profits as if it would continue indefinitely with a 5 percent interest rate. Investors recognized the favorable terms by bidding up the price of telegraph stock while other stocks were generally dropping. Electric Telegraph went from 132 in 1867 to 270 in 1870 (the date of nationalization), while the Magnetic rose from 90 to 170 and the UKTC from 1.5 to 6.5 during the same period.[5]

It is hardly surprising that the nationalized system failed to live up to the claims made for it by its supporters. The government fully capitalized the potential monopoly profits by paying twenty times the cartel profits for the system. It then reduced the charges to 1s. anywhere in the United Kingdom, a price lower than that charged under the price competition prior to the cartel. Large expenditures were incurred in putting the telegraph system into proper working order because maintenance had been deferred during the pending nationalization. Additional capital was employed to extend the system to places the cartel had not found profitable to serve. Only if there had been gross inefficiency or a severe underestimate of the demand elasticity under private operation could the government have expected to break even on its telegraph operations. The takeover was financed by consols paying 3 percent interest. In the first decade of government operation, the telegraph revenues were adequate to pay operating expenses and interest charges only in 1870, the year of nationalization.

There is some question as to why the Post Office sought the takeover even at an unreasonably high price. We would expect the Post Office to desire control of the telegraph which provides potential competition to the mail. In the United States the Post Office supported government purchase of the original Morse patents, even though other branches of the government did not. However, we also expect a government agency to minimize noise or discontent with its operations. The British Post Office had been held in very high regard before the telegraph purchase, but its inability to cover the interest charges on telegraph debt from telegraph operations led to extensive criticism and official investigations of the efficiency of Post Office operation of the telegraph. While the nationalization provided an extension of power and scope of operations for the Post Office, it also reduced the public image of the Post Office and brought its operations under greater official scrutiny. It is possible that Post Office officials really believed that the integration of the telegraph and mail operations would provide enough of an increase in efficiency that the Post Office could afford to pay a monopoly price and still cover expenses with reduced charges. Alternatively the officials may have been willing to accept criticism for telegraph losses in order to achieve the benefits of unified control and explicit legal monopoly over the nation's communication operations.

While the Post Office was criticized for its losses on the telegraph system, it was praised for its efforts at improving the service. The uniform 1s. rate for messages anywhere in the United Kingdom, together with expansion of the telegraph system to areas not previously served, greatly stimulated demand. The number of messages sent was multiplied by a factor of four between 1869 and 1879. United States messages increased threefold during the same time period under the profitable Western Union monopoly.[6]

The Telephone in Great Britain

Although the early British telegraph development was based on English inventions, the telephone development was based on the American inventions of Bell and Edison. In February 1878 an agent for the Bell company began negotiations with the British Post Office over rights to establish a telephone system. The Bell company first attempted to sell its patent rights to the Post Office, as it had attempted to sell its United States patent rights to Western Union. No immediate agreement was reached because the Post Office was unable to secure the necessary authorization to proceed from other branches of the government. The Bell company formed a private British company in June 1878 and began building telephone exchanges. The next year the Edison Telephone Company was formed with rights to the Western Union–owned Edison patents and the two companies began British competition similar to that between their parent companies in the United States.

The activities of the two companies caused the Post Office to assert its monopoly rights over telephones. Whether or not the monopoly existed was a disputed legal question based on the interpretation of the telegraph monopoly act. The Post Office offered to license the telephone companies to operate under restrictive terms. The companies would be required to pay a fixed fee plus 25 percent of gross profits to the Post Office, and would be limited to providing service within a half mile of the central office. The companies would be prohibited from providing connections between any two exchanges. The terms offered by the Post Office were designed to protect its telegraph system for which it had paid a high price only nine years earlier. The restrictions placed on the companies would prevent them from building long-distance lines to compete with the telegraph and would allow the Post Office to gain a share of the local exchange profits from the new invention. However, the Post Office had no patents to offer and was basing its position on its arguable status as a legal monopolist of the telecommunications services. The telephone companies asserted that the monopoly law did not apply to telephone systems and continued to operate without requesting licenses from the Post Office. The Post Office then filed suit against the companies at the end of 1879. While the suit was in progress, the two telephone companies merged to form the United Telephone Company following the precedent set by the Bell–Western Union settlement in the United States.

At the end of 1880 the court found in favor of the Post Office and gave a broad interpretation to the telegraph monopoly grant which applied the monopoly to existing and future methods of communicating by electricity. The Post Office then proposed to set up its own telephone system but was rebuffed by other agencies of the government. The inability of the Post Office to cover expenses on its telegraph operations made other agencies of the government skeptical of the wisdom of having the Post Office control the telephones despite its legal right to do so. The Post Office was required to proceed with negotiations to license private companies to provide telephone service and was given the right to establish a limited telephone system of its own to improve its bargaining position.

In the spring of 1881 the Post Office and the telephone company reached agreement on the terms for telephone development. United Telephone agreed to dismiss its appeal of the monopoly ruling and accept the necessity of a Post Office license in order to operate. It also agreed to grant the Post Office a patent license so that the Post Office could enter the telephone business in competition with it. In addition, the company agreed to pay the Post Office 10 percent of gross receipts and to allow the Post Office to erect all long-distance telephone wires.[7] In return, the Post Office agreed to allow United Telephone to exist.

United granted far more concessions to the Post Office than Bell did to Western Union in spite of the fact that Western Union had both market power and important patents, while the Post Office had "only" a court decision that it was entitled to a monopoly of telephone service. However, given the monopoly ruling, United Telephone's patents were of no value at all. Its only option was to accept the Post Office terms or go out of business. While it might have been able to prevent Post Office use of the telephone because of its patent rights, it would not have been able to develop the phone itself because of the monopoly ruling. Its only real bargaining chip was the possibility of a reversal of the monopoly ruling if the appeal were pursued. So long as United Telephone believed that the probability of a reversal was low, and that the potential profits from exchange service alone were high enough to pay the Post Office royalty and still make a better than competitive return, it was in its interests to accept the Post Office offer.

The telephone company continued its development of telephone exchanges under the restrictive Post Office license. Rates were set very high. The Post Office did not build any long-distance lines but left all long-distance business to the telegraph. The high prices, limited number of subscribers, and inability to communicate by telephone with nearby exchanges all combined to reduce demand for the service. Telephone development proceeded slowly and was restricted to the center of the largest cities. The inability to provide long-distance service and uncertainty over future Post Office action reduced the company's ability to plan for long-term exploitation of the telephone technology and induced it to gain maximum short-term profits.

Following agitation against the Post Office over the restricted state of telephone development, the Post Office amended the contracts in 1884 to allow the telephone company to provide long-distance oral service, while reserving written message service to the Post Office. About the same time the Post Office reduced the telegraph rate from 1s. for a message of twenty words to 6d. for a message of twelve words, almost a 50 percent reduction in the average cost per telegraph message. The price cut stimulated telegraph traffic and reduced the economic incentive and the popular pressure to establish long-distance telephone service. The price cut reduced the telegraph revenue below the level of operating expenses without any charges for interest on the debt. Despite the below-cost telegraph competition, the telephone company did build some short long-distance lines, but the long-distance network remained in a primitive state relative to the long-distance network in the United States.

When the British patents expired in 1891, the Post Office rescinded the right of the telephone company to build long-distance lines. After that date, long-distance telephone lines were operated by the Post Office at high rates. Only for calls of fewer than twenty miles was the charge as low as for a telegraph message. For 400-mile calls such as London–Glasgow, the charge was 5s. 6d. compared to 6d. for a telegraph message. The Post Office agreed to allow competition in local exchange service and licensed additional telephone companies or municipalities to build local exchanges in return for a 10 percent royalty to the Post Office. Competition brought a significant reduction in rates (sometimes to as low as one quarter of the monopoly rate) and greatly stimulated demand. In London, the Post Office itself built a competing exchange.[8]

The Post Office control of all long-distance lines, plus its competitive London exchange, gave it an important operating role in the telephone service. It also began buying up local municipal exchanges that had run into financial difficulties. After a short period of direct competition in London, the Post Office and the company reached an agreement to provide complete interconnection and to charge the same rates. The agreement also provided that the government would purchase the company in 1911, at the expiration of its license from the Post Office. At that time the telephone system was brought under direct Post Office control.[9]

By prohibiting private development during the first period, allowing it under a short-term license during the second period, and engaging in its own development with high prices relative to the telegraph during the final period, the Post Office extended the life of telegraph technology relative to long-distance telephone technology. The British long-distance network remained very undeveloped at the time of nationalization in 1911, while in the United States Bell was straining to find a way of providing coast-to-coast conversations. The number of telegraph messages almost tripled in Britain between 1885 and 1900 while the number in the United States only increased by 50 percent during the same period.[10] Yet telephone technology could have been expected to supersede telegraph earlier in Britain than in the United States under the same economic conditions because Britain's compact size made it technologically feasible to communicate anywhere in the country by telephone much earlier than in the United States.

The Telephone in France

In France, the optical telegraph attained its greatest development. Beginning in 1792, military communications were carried on through stations located at six- to twelve-mile intervals equipped with signal arms that were viewed through telescopes from the next station. The lines of optical telegraph were greatly extended during the Napoleonic wars, and by 1842 over three thousand miles of optical telegraphs were operated by the French War Department for government use only.[11] Following an attempt to establish a commercial optical telegraph line between Paris and Rouen, a law was passed in 1837 that declared the telegraph a government monopoly. Thus the French policy toward the electric telegraph was in large part determined even before its invention.

Morse visited France and attempted to secure permission to establish a commercial telegraph system without success. The building of a short railroad telegraph line between Versailles and St. Germain led to an explicit ban on further private construction. The French government built its first line in 1845 for government use only. Development proceeded slowly because the government was reluctant to abandon its system of optical telegraphs. Following the change of government of 1851, France began to seriously develop an electrical telegraph system. The system was primarily for government use and under the control of the head of the national police until 1878, but it was also open to the public when not needed for official business. During the time preceding the invention of the telephone, the French telegraph development was controlled by military and other government needs and not by commercial demand.

The telephone was first exhibited in Paris in 1878. The next year American representatives of the competing phone systems began seeking French business. The government telegraph engineers were skeptical of the merits of the telephone and unwilling to adopt it. The telegraph authorities decided to allow private telephone development under a restrictive government license. The government offered a five-year license, without a monopoly guarantee, in which the telephone system would be built by government engineers and paid for by the private company. The company would pay a royalty of 10 percent of gross receipts to the government. At the end of the five-year period, the government would have the right to purchase the system from the company and enter the business itself without payment for patent rights. A. N. Holcombe, a historian of the early Continental telephone development, summarized the effect of the French franchise as follows:

> The conditions could not be regarded as favorable to the growth of the telephone industry . . . Their whole attitude seemed to have been dictated by the one purpose to evade the risk of introducing the new service by shifting the responsibility to private promoters. At the same time they protected themselves against unexpected success by making the term of the franchise short. Consequently, in order that private enterprise might be induced to undertake the risk at all, it had to be allowed liberty to recoup its advances quickly from the consumers.[12]

During 1879 three separate systems were licensed, based on the Bell, Edison, and Gower patents. All three companies were authorized to operate in Paris, while two were authorized for other major French cities. The three companies merged together in 1880, following the American and British examples. The Paris exchange was opened in 1881 and service to other major cities followed, but no efforts were made to establish service in the smaller cities. In 1884 the company franchise was extended for another five years, but the phone system continued to develop slowly. Uncertainty regarding its ability to continue in business beyond 1889 or the compensation to be paid in a government takeover induced the company to pursue short-run profits rather than long-run system development. By 1888 the company had a total of 8,459 subscribers of whom 6,120 were in Paris. The entire French telephone system was far smaller than the system Western Union had turned over to Bell after the short period of early competition in the United States.

The development of long-distance service proceeded even more slowly than the development of local service. No lines were built during the first five-year franchise because the private company lacked authorization and the telegraph authority did not wish to. After 1885 the government began building long-distance lines and completed a Paris–Brussels link (200 miles) in 1887. The resulting drop in telegraph traffic aroused the government's concern to protect its telegraph revenues and retarded further long-distance telephone construction.

By the time the second five-year franchise expired in 1889, there was considerable dissatisfaction with the state of French telephone service. French telephone rates were among the highest in the world. Service was limited even in Paris and practically nonexistent in the rest of the country. Holcombe described the predicament of the French telephone system as follows:[13]

> The company feared purchase in 1889, and hence would not establish new exchange systems, improve those already in existence, or reduce its rates . . . Nor could the government construct new works itself on account of the difficulty of securing appropriations from Parliament. Parliament was unwilling to appropriate the public money until it knew what the ultimate disposition of the plant was to be, and in fact had not appropriated anything for fresh construction since its first appropriation in 1882.

Various plans were proposed for continuing the partnership of government and private enterprise, but no solution could be found which satisfied all interests. The government then passed a bill that nationalized the telephone system in 1889. The company refused to surrender its property peaceably because of an earlier agreement to extend its franchise. The government took possession of the telephone system by force and left the appropriate compensation to later legal proceedings.

The government take over did little to improve the French system. The financial difficulties of the French government prevented it from making significant appropriations to build the system. The inability of the government to finance construction led to a cooperative plan in which people who desired service would raise the necessary construction capital and loan it to the government without interest. The loan would be repaid out of profits if the exchange was successful. If the exchange was unsuccessful, the people who loaned the construction funds would bear the loss. The arrangement freed the government from the responsibility to raise capital or bear risk, but allowed the government to retain ownership of successful exchanges.

Consumers accepted the system because of the great demand for telephone service and the legal inability to satisfy the demand through private initiative, but it could only work in a time of great disequilibrium between real costs and demand. The potential subscribers bore all the risk but could receive none of the benefits of a speculative enterprise. In effect, the consumers were required to build the exchange and give it to the government without charge because the loans would only be repaid out of the profits from the rates the consumers paid for the service. Yet they had no guarantee that the rates would be set at the level of operating costs once the loans were repaid, or that the government would use the future profits to improve and maintain the exchange. In fact, the exchanges were often not maintained properly once they were in operation, and the lack of capital continued to hamper French telephone service.[14]

Long-distance lines were financed in a similar manner to local exchange lines. People who expected to benefit from the use of the line advanced money to the government to cover construction costs. The government was responsible for maintenance and operation, but maintenance was often deferred for lack of adequate funds. A study of service on the Paris–Marseilles line for the year 1906 showed a total of 204 service interruptions lasting an average duration of 14.5 hours each.[15] Long-distance service was limited in coverage and unreliable. The deficiencies in the long-distance service were caused by inadequate capital availability to the telegraph authority as well as the attempt to protect telegraph revenues and capital from telephone competition.

The inability of the French government to finance telephone development prohibited telephones from even making the maximum contribution to the public treasury. Rates were kept above the monopoly profit maximizing rate, not because the authorities believed there was no elasticity of demand, but because they believed there was substantial elasticity and were unable to build facilities to serve those who would desire service at lower rates.[16] Limiting investment capital to funds received from operating profits or zero-interest private loans forced the system to grow slowly and prevented the government from developing either a profit maximizing or a socially optimal telephone system.

The Telephone in Germany

An experimental military telegraph line was built between Berlin and Potsdam in 1846. The Prussian government was satisfied that the electric telegraph was better than the old optical telegraphs for military operations and extended the system to other cities. In August 1849 the system was opened to public use when not needed by the government. Administration of the system was later shifted to the Post Office. Although the system was begun to meet military needs, the public demand for telegraph service was allowed to exert a greater influence on system development than in France.

The head of the German Post Office was very receptive to the invention of the telephone. Rather than seeing it as a competitor to telegraph revenues, he viewed it as a device to extend telegraph service to places that were too small to support a telegraph office. Because the telephone did not require a skilled operator, a public telephone could be placed in a village with a line to the nearest telegraph office so that telegraph messages could be called in. During 1877, contemporaneous with the earliest Bell telephone sales in the United States, the German Post Office began installing telephones in small villages to connect them to the telegraph network. Fifteen villages were connected in 1877 and another 272 in 1878. Because the German village connections were begun while the United States had only private-line telephones, it is sometimes asserted that Germany had the first public telephone network in the world.

The Post Office did not begin to establish urban exchange service until American agents requested permission to establish private systems. In Stuttgart a representative for Bell secured municipal but not state permission to build a telephone exchange. He began construction in 1880 but was stopped by the state police. The Post Office then declared an official state monopoly of telephone service and began the construction of its own telephone exchanges. Telephone service was established in Berlin, Hamburg, Cologne, and other major cities in 1881. The first long-distance service was instituted in 1884 and developed slowly. Long-distance telephone lines were built only where demand exceeded the capacity of existing telegraph facilities. Consequently the telephone did not displace existing telegraph investment but was allowed to displace additions to telegraph capacity.

Before a local telephone exchange was built, the community was required to collect long-term service commitments from enough people to guarantee the economic success of the service. Similarly, before long-distance lines were constructed, the local communities were required to guarantee that the revenues from the line would be adequate to cover its costs. The financing method was not so restrictive as the French system because the subscribers did not have to actually advance the money, but it imparted a conservative bias that prevented rapid development. People were required to commit themselves to use of the local exchange or long-distance line without being sure with whom they would be able to communicate. They would thus have a tendency to underestimate their actual demand. If the line were built, additional people could order service with less risk than was incurred by the initial subscribers. However, if everyone waited until the exchange was built to order service, the Post Office would not build it.

The Post Office initially charged a single rate for exchange service in all cities served. The rate was relatively high (beginning at 200 marks per year and reduced to 150 marks in 1884), and allowed substantial profits to the telegraph administration. Political agitation began for reduced rates in general and especially for reduced rates in the smaller cities. The smaller cities were relatively undeveloped at the original rate, because the service was of less value in the smaller cities, which had shorter distances and smaller numbers of people with whom to connect. The cost of service was also less in the smaller places because of the smaller switchboards and the shorter length of wires from subscribers to the central office. By 1885 there was almost no city exchange service in German cities of less than fifty thousand.

Although the inadequacy of the flat rates was widely recognized, concern for protecting existing rights prevented change. The telegraph administration was unwilling to accept lower average rates. Users in large cities were unwilling to see their rates raised in order to benefit the actual and potential users in the smaller cities. Representation of both sets of interests in the Reichstag resulted in a deadlock and perpetuated the 1884 rates until 1900. The protection of property rights in the status quo prevented service to smaller cities. It appears that the telegraph administration's fear of lower average rates was groundless. The existing rates restricted service to large cities and were profitable for that class of service. Lower rates were sufficient to cover costs in smaller cities. Thus no reduction in total profits would have come from maintaining existing rates in large cities and instituting lower rates in smaller cities.

A new rate plan that made allowance for city size was finally instituted in 1900. The new rates left the charge at the long-established 150 marks/year in medium-sized exchanges. Charges were increased to a maximum of 180 marks/year in the Berlin exchange and reduced to a minimum of 80 marks/year for exchanges with under fifty telephones. A measured service was also instituted with a reduced yearly charge and a charge for each call. The new rates stimulated a rapid expansion of phone service, especially in the smaller cities. The total number of telephones increased 150 percent between 1900 and 1906, from less than 200,000 to over 500,000. The number of telephones in cities with populations between 2,000 and 5,000 increased tenfold in the same period.[17]

The Telephone in Sweden

The Swedish government operated the telegraph except for railroad lines, but it neither began telephone development nor limited the initial telephone activities of private companies. In contrast to the rest of Europe, the Bell representatives were able to establish telephone service without a restrictive government license. However, Bell was not granted a patent monopoly in Sweden. International Bell established exchanges in major cities while people in rural areas set up mutual telephone systems.

In 1883 the General Telephone Company was formed to compete with Bell in Stockholm and cut the Bell rates about 30 percent. Bell made no response to the General threat and retained its old rates. By the end of 1884 General's system was almost three times the size of Bell's.[18] After Bell lost its dominant position to General, the two companies worked out an interconnection agreement to give the subscribers to either system access to the phones of the other. In 1890 General acquired a controlling interest in the Bell Company.

When General began connecting towns around Stockholm with its system in 1884, the state telegraph agency recognized that telephone service was a threat to telegraph revenues. In a law that protected the existing General position and the existing telegraph monopoly, the telegraph agency was awarded monopoly rights to all long-distance telephone wires except within seventy kilometers of the center of Stockholm, while the private companies were given freedom within the Stockholm area. The telegraph agency then announced that the use of government long-distance lines would be restricted to subscribers to government exchanges. Thus the government monopoly over long-distance lines (enforced by legal right) was to be used to develop a monopoly of local exchanges through competitive pressure. The small companies that had established systems in individual Swedish cities sold to the state, but the General Telephone Company resisted takeover. The government established competing local exchange service within Stockholm using reduced rates and the long-distance privileges as incentives to gain new subscribers.

In 1891 General Telephone agreed to sell all lines over seventy kilometers from Stockholm to the government in exchange for rights for General customers to use the government long-distance lines. Extensive competition between the government exchange and General followed the agreement. The companies cut rates and introduced a variety of technological improvements, including the use of all metallic circuits to improve the quality of the conversations. Several new classes of service were introduced. These included two- and four-party lines, and a minimal flat payment plus a message charge to induce infrequent users to join the system. The total effect of the rate wars was to reduce the original Bell rate of 160–280 kronen per year to a rate of 60 kronen for business and 50 kronen for residential service. The rates went as low as 10 kronen ($2.65) minimum charge per year for a phone with message charges for low-volume users. Holcombe summarized the effects of competition as follows:

> These rates were lower in Stockholm than in any government exchange outside of the area subject to competition. In other words, the government was discriminating between its own citizens on the sole ground that the lower rates were needed for competitive purposes . . . This prolonged rate-war, combined with an excellent service, quickly produced a telephone development in Stockholm that put it far ahead of any other city in the world.[19]

The government agency finally won the competitive battle by repudiating its agreement to allow General customers access to government long-distance lines. Isolated by the cutoff from government lines and its legal inability to build its own lines, the General company merged into the government agency and completed a government monopoly of the Swedish telephone system.

European and American Comparison

The invention of the telephone in the United States and the early competition between Western Union and Bell gave the United States an early lead in telephone development. By 1885 the United States had one telephone per 500 persons, a far higher rate than the one per 1,000 persons in Sweden, one per 3,000 in Germany, and one per 5,000 in France. By 1895, after ten years of slow growth under the Bell monopoly, the American development of one phone per 235 persons was only half of the competition induced Swedish rate of one phone per 115 persons. The American rate was still well above the German rate of one phone per 397 persons and the French rate of one phone per 1,216 persons. The intensive American development under competition between 1895 and 1906 put the United States far out in front. In that year, the United States had one phone for every 15 persons while Sweden led the European countries with one phone for every 39. Stockholm itself continued to be the world's best telephone-equipped city with one phone for each seven people, well above the rate of major American cities. France continued to trail at one phone for 244 people.

The significance of competition in the United States is even more apparent in the figures for different parts of the country than in the aggregate figures. The government agencies concentrated their efforts in the major cities, and especially in the nation's capital. Little attention was given to smaller towns and rural areas. Similarly the Bell monopoly concentrated its exchanges in major American cities. In contrast, the American competitors put most of their effort into developing the rural areas and small towns and were less successful in competing with Bell in the major cities. By 1902 Berlin and New York City had almost equal telephone coverage with one phone per 39 persons in New York and one per 43 in Berlin. At that time, Iowa had twice the development of New York City with one phone per 19 persons provided by a total of 170 different systems. Rural districts in Germany had less than one-tenth the development of Berlin with one phone per 500 people.[20]

The rate of telephone development under competition was more rapid than under private monopoly, while development under private monopoly was more rapid than under government monopoly. With competition, low prices and the striving for competitive advantage induced rapid development. The existence of several companies allowed various beliefs as to the elasticity of demand to be tested and prevented slow growth through a mistaken belief that the demand was inelastic. Private monopoly will in general charge higher prices than competitive companies, and will therefore have a lower level of equilibrium demand, but it must attempt to satisfy the demand in order to maximize profits and reduce the incentive for another company to enter the business. The monopoly will generally be able to finance rapid development so long as it is profitable.

A public monopoly with a legal barrier to entry may on occasion develop its field rapidly with low prices and good service, but there are no economic forces to require it to do so. If the public agency is unable to float its own securities so that it depends on public appropriations for capital, it is likely to encounter difficulty gaining enough capital for rapid growth regardless of the profitability of its operations. It may be forced to expand only through reinvesting operating profits (as in the case of French telephones) and thus be limited to a growth rate equal to its profit rate. It lacks the private monopolist's incentive to satisfy demand in order to prevent competition, but may make every effort to satisfy existing demand in order to avoid political criticism. If public capital for rapid growth is limited, the agency may set prices above the profit-maximizing level in order to generate high profits for reinvestment and to choke off excess demand. Political criticism over high rates (with a good standard of comparison often lacking) is likely to be less severe than criticism over the agency's inability to serve those who desire service at the existing rates.

Market power in Europe came primarily from government franchise. Even patents played a relatively minor role. The government agencies used similar tactics to those of private companies to extend their power when it was not firmly established by law. The Swedish agency used price warfare and systems advantages of its protected long-distance system to extend its monopoly to local service in Stockholm. In France and Germany, there was no need for government agencies to develop strategies to extend their power because they were granted full authority over development by the state.

The government agencies used their market power to protect themselves from risk rather than to maximize profits. Slow development and high rates guaranteed that the government would not find itself with unusable capital investment. The French agency provides the most extreme example in its practice of requiring customers to supply the capital to build their system, but other government agencies also followed risk averting practices. With absolute barriers to entry, the government could develop the system in a conservative manner without inducing less risk averse firms to come into the market. It is likely that the initial slow development under private monopoly in British telegraph and American telephones were also related to risk aversion. In both cases, the private monopolist apparently developed the system more slowly than the maximum profit rate, but in so doing reduced the risk to the company. The potential risks included the possibility of unproductive capital investment if demand were less than expected and the possibility of loss of control of the firm to outside financiers from the need to raise large amounts of capital. All the monopolists examined developed their systems primarily from internally generated capital. The capital was derived from profits, customer advances, or from privately placed stock to existing stockholders. Little reliance was placed on commercial loans, publicly placed stock, or government appropriations. The conservative internal financing used by both public and private monopolists made high prices and slow growth appear attractive even if it was not the route to maximum profits.

The risk-averting behavior of the public monopolists with regard to system development and financing was continued in their reactions to a major new technology. All the public agencies attempted to protect their telegraph services from telephone competition, even when they controlled both telegraph and telephone. The response was not merely a monopolist maximizing profits; their control of both modes would have allowed greater profits by more extensive telephone development. It is better viewed as a risk-minimizing response. The telegraph was a known technology with substantial existing investment. Telephones were a new technology with unknown demand and uncertain operating characteristics. Rapid development of long-distance telephone would have caused telegraph capital losses if it had been very successful and potential telephone capital losses if it had been unsuccessful. Either result could cause political problems for the government agency. The conservative response of protecting telegraph investment while experimenting with slow telephone development left critics only with rather vague assertions that the agency should do better.

The long-distance development of Europe provided the most contrast with the United States. The Bell–Western Union agreement placed telephone and telegraph technology under separate control and provided Bell with strong incentives to develop long-distance technology in order to free itself from dependence on Western Union. Long-distance development took place in the United States without concern for protecting telegraph physical capital or the human capital of telegraph employees and engineers. Both were protected in the European countries by placing long-distance telephone under the control of the telegraph agency. In the United States, long-distance technology was pursued as a source of post-patent market power. In Europe, long-distance technology was held hostage to established telegraph interests.

The actions of the European government agencies were consistent with the regulatory response of maintaining property rights in the status quo discussed in Chapter 1. A technological innovation was not allowed to break down established market power, capital values, rate structures, or operating procedures developed in the telegraph. The changes came in slow, measured steps rather than the abrupt shocks to the status quo that are induced by market forces during technological revolutions. Change was not totally prohibited as it could have been by the government monopolies. Long-distance telephone networks were eventually built, but the new technology was required to "buy out" the old through license fees and restricted development. The initial French attempt to restrict the development of the electric telegraph to protect its investment in clearly inferior optical telegraphs provides the most extreme example, but all the government agencies restricted new technology to some extent to protect established rights in existing methods.

6 The Decline of Competition, 1907–1935

The Management Change

By 1907 AT&T's market power was threatened. The company was still the dominant firm, but it had been losing market share steadily. Its main advantage over its competitors was its long-distance network, which allowed Bell subscribers a wider range of communication than that available to non-Bell subscribers. Although the company possessed important patents on long-distance technology, it could not be certain of its ability to prevent entry into long distance. The large number of isolated independent operating companies offered a strong incentive for the building of a competitive long-distance network that could weld the competitors into a unified opponent of the Bell system. The development of a competitive long-distance system would transform the industry into a duopoly rather than a dominant firm industry and make Bell's ability to earn monopoly profits dependent on successful cooperation with the other firm. Even without a competitive long-distance firm, Bell's pricing freedom was constrained by the presence of local competitors who were willing to expand at Bell's expense.

Another threat to Bell's market power was the invention of radio. Bell's research program had been oriented to telephone-related devices and not to general technological advance. Although the company had been willing to purchase relevant outside technology, such as the Pupin loading patents, it had not acquired an interest in completely new technologies such as radio. Radio was coming into common use for ship-to-shore telegraphic signaling in 1907, but was incapable of carrying voice signals. It was therefore not a direct competitor to telephones, but represented the kind of major technological change that could displace Bell as the dominant telecommunications firm if it were developed outside of Bell control and influence.

An additional problem was the possibility of government regulation or nationalization. While neither would necessarily be against the interests of the firm, either regulation or nationalization could be harmful. The city franchise regulation which had existed during the competitive era had been helpful to Bell because it imposed greater costs and burdens on the new competitors than on the established Bell company. Popular support for regulation was increasing, forcing the company to decide whether to support regulation or oppose it. If the company supported regulation, it could influence the form of regulation. If it opposed regulation, it risked having an unfriendly form of regulation forced on it.

During the monopoly period, the company had financed growth internally or by placing new securities with existing stockholders at prices less than the market value. Because the stock was widely distributed, there was no dominant stockholder group to impose its will on management. So long as profits were high, the stockholders had no reason to question the efficiency of management. However, the coming of competition indicated that the monopoly-period management had not been entirely efficient. A protected company tends to develop slack or inefficient practices.[1] The absence of perfect information and perfect foresight means that any managerial team will make mistakes. In a competitive market, those mistakes are revealed by the success of other firms that are either more skillful or more fortunate. However, in a monopoly market, there is no real check on whether or not top management's decisions are the best possible. Thus the Bell monopoly was able to charge prices above the profit-maximizing level, presumably without being aware of it, because the demand was not tested at lower prices until the competitive era. The fact that expenses per telephone increased steadily during the monopoly era and then decreased steadily during the competitive era suggests that management was not at maximum efficiency during the monopoly period.[2]

The rapid growth and reduced profitability during the competitive period greatly increased Bell's need for outside capital. As Bell's need for outside capital increased, the constraints on managerial freedom also increased. The company was brought into regular interaction with New York banks and financiers and was forced to be responsive to the views of new investors as to how the company should be run. The success of financial promoters in forming U.S. Steel and other major corporations by merging together essentially all the companies in an industry caused them to question the wisdom of Bell's continuation of competition rather than arranging mergers with its competitors. It appeared that greater profits were available in the telecommunications industry through consolidation of competing companies than with the existing structure.

In 1902 AT&T sold a block of 50,000 shares to New York banking interests on the condition that John Waterbury, president of Manhattan Trust Company, be elected as director of the company.[3] Three years later, director Waterbury suggested a long-term financing plan to satisfy the capital needs of the company for some time to come. The essence of the plan was to sell between $85 and $135 million worth of convertible bonds to J. P. Morgan and associates. This proposal was viewed with concern by some of the officers of AT&T as shown by the following letter from the company's counsel, treasurer, and vice-president to the president:

> To our minds there is another risk in the proposed plan which should be had in mind. If a bankers' syndicate should be formed, under the proposed plan, who should pool their bonds or place them in trust, the trust so formed, by exercising the option given for the conversion of bonds, would have the power to acquire so near an absolute controlling interest in this company as practically to control the whole assets of the company, which they could use for any schemes of financing that they saw fit. In short, having nearly one-half of the entire issue of capital stock of the company, they could consolidate this company with other companies, or make any other arrangement in regard to its future financing that they saw fit.[4]

The company subsequently rejected the plan and made other arrangements for the necessary financing that year. However, the proposal was revived later in the year, this time for $150 million worth of convertible bonds, and was approved by the officers of the company.

The bankers did not find the expected good market for the bonds. They were not offered publicly at all during 1906 because of depressed bond market conditions that year. When the public sale began, only $10 million of the bonds were purchased, leaving $90 million of the initial $100 million offering in the hands of the underwriters. It is unclear whether the inability of the bankers to sell the bonds caused them to take an active role in management, or whether the bankers preferred to hold the bonds themselves in order to have a management influence; but, in any case, Waterbury suggested in early 1907 that a special committee be appointed to consider the management and organization of the company.

The review committee was made up of outside directors chosen by the banking interests. One of the "bankers' directors" and review committee members was Theodore Vail, who had resigned in 1887 and pursued other business interests. As a result of the review, the president and some of the other officers resigned and Vail was selected as president.[5] With the support of J. P. Morgan and other New York financial interests, Vail developed a three-pronged strategy to restore AT&T's market power and profitability. The strategy included: (1) merger with a telegraph company and the independent telephone companies, (2) a welcome to regulation, and (3) increased emphasis on fundamental research and the purchase of important outside patents.

Mergers and Antitrust

The price competition between 1894 and 1907 had not been severe enough to bankrupt most of the independents, but it had kept their profit rates modest and made them more open to an attractive merger proposal than they would have been earlier in the competitive era. The competitive period had also removed some of the emotional desire to fight the Bell monopoly and had focused the stockholders' attention on the profitability of their investment. The almost complete coverage of the country with telephone systems and the reluctance of cities and states to grant franchises to additional companies made the merger policy less likely to induce new entrants than it would have if undertaken at the beginning of the competitive period.

Because Bell's advantage over the independents was increasing with the increasing importance of the long-distance lines, the company could possibly have purchased its competitors at lower prices if it had pursued a competitive policy for a longer time period. However, that strategy would have reduced short-run profits and increased the risk that a competitive long-distance system could be established. Although Bell held the Pupin patents that gave it substantial advantages in long-distance operation, the existence of Western Union and Postal Telegraph with extensive wire lines and the existence of many independent telephone companies left open the possibility that a method could be found to effectively merge the competitors together. A policy of telegraph and telephone acquisition would reduce the risk of long-distance competition and increase short-run profits.

Vail first chose Postal Telegraph as a telegraph merger partner because of its international cable business. He wrote in 1906:

> From the very beginning of the "Telephone" business, so far as I have had to do with the policy of the Co., it was directed toward the ultimate absorption of the "Telegraph" business—I do not remember that I was alone in this, and as I believe and understand, this policy still exists.
>
> Any fight over the domestic telegraph business would result in disaster to the new earnings of the "Western Union" while it is doubtful if it would be particularly noticeable in the makeup of the balance sheet of the Mackay "Cos." . . .
>
> . . . It would be good policy to acquire the Postal system, if it could be got as I believed it could at a cost which was fully represented by useful property, utilizing the organization to carry on the telegraph business, and also use it to handle the opposition telephone business.[6]

The head of the Postal Telegraph, Clarence Mackay, was also interested in a merger but he wanted to control the combination. Mackay began purchasing AT&T stock on the open market and became the largest single stockholder (with 5 percent of the stock in 1907, four times the holdings of the second largest stockholder) but was denied representation on the AT&T board of directors.

The aggressive Mackay was unable to gain control of AT&T and unwilling to have his company absorbed, causing Vail to look to Western Union for a merger partner. Western Union had been suffering from managerial difficulties as well as competition from Postal and AT&T's long-distance network. The stockholders were therefore quite willing to turn control of the company over to AT&T. In 1909 AT&T purchased 30 percent of Western Union's stock at a premium of about 13 percent over the market price and received additional voting rights to give it working control. Vail was chosen president of Western Union as well as AT&T and seven common directors of the two companies were chosen.[7] The common president allowed the companies to be operated in a unified manner as if a complete merger had taken place. The merger agreement brought telephone and telegraph together under one company thirty years after the telecommunications industry had been divided into the telephone and telegraph parts by the 1879 Bell–Western Union agreement.

AT&T also began to purchase independent telephone companies. The attempt to buy independents was opposed by some and welcomed by others. As MacMeal described the situation in 1908:

> Most Independents still felt very intensely over what they described as the "Bell taint" and bitterly fought Bell purchases of Independent properties and Bell wire connections. This, of course, did not deter the Bell people from buying any plant they desired and could get. And they picked up a good many of them, due to different reasons. Sometimes the owners had failed to make the profits they had believed were in the business; sometimes they were offered such a handsome bonus on their investment that they could not resist; in other cases, cut rate warfare exhausted their resources and forced a sale.[8]

Purchases were undertaken directly by AT&T or its operating companies and also by various agents including J. P. Morgan and Company. There was some confusion and uncertainty among the independents regarding the final plans for the property purchased by Morgan. After a major purchase of independent properties in Ohio by Morgan, MacMeal stated that many people believed they were purchased on behalf of Bell but that also "it was reported that the Morgan group planned to develop and connect the properties to compete with the Bell."[9] In later discussions over the connection between Morgan and Bell, MacMeal reported: "At a hearing in Washington, H. P. Davison, Morgan partner, declared on oath that there was no hook-up between the Morgan purchases and the Bell organization."[10] The question of whether or not the Morgan purchases were on behalf of Bell was significant for the independents' selling strategy. If the Morgan purchases were for Bell, the price offered for additional systems could be expected to decline over time as the remaining independents became more and more isolated and thus easier to drive out of business through competition. However, if Morgan were planning to establish a competitive system, the prices for remaining independents could be expected to rise over time because of offers from both systems. Thus the question was important to determining how anxious they should be to accept a first offer for merger.

In 1911 an attempt was made to consolidate the industry at once rather than by individual purchases. A committee was chosen from the independents to negotiate with representatives of Morgan and Bell. However, the negotiations broke down because of disagreements among the independents.[11] Further acquisitions were made on a company-by-company basis.

The merger policy reversed the previous steady decline in Bell's market share. The independents increased their total telephones controlled by 22 percent from 1907 to 1912 but Bell increased its telephones by 62 percent during the same period, reducing the independents' market share from 49 to 42 percent. The Bell system growth was much slower during the period of mergers, dropping from an average of 21.5 percent per year during 1895–1906 to an average of 9.6 percent per year during 1907–1912. This was due to a stabilization in the price of service (the average revenue per telephone dropped slightly and irregularly during 1907–1912 compared to a sharp and steady drop in the previous period) and the absence of significant areas totally without service to provide rapid growth in the system.[12]

In some cases, the acquisitions were detrimental to the remaining independents. For example, the Home Telephone Company of Clarksville, Tennessee, was connected with other independents through the Long Distance Telephone and Telegraph Company. When the latter company was purchased by the Bell operating company, connections to the Home Telephone Company were cut off. In another case, the Memphis Telephone Company and Tri-State Company had a long-distance line connecting them. The Tri-State Company was purchased by Bell and service on the connecting line was stopped. Many companies filed antitrust charges against Bell either to prevent the completion of mergers with potentially harmful effects or to seek court protection for their connecting rights. Actions under the state antitrust laws by private companies and state attorneys general prevented some mergers and established limitations on Bell's post-merger conduct toward remaining independents.[13]

While the state antitrust actions had placed some restrictions on Bell's conduct, the company was much more concerned about the possibility of federal antitrust action under the Sherman Act. The successful prosecution of Standard Oil and American Tobacco had proved that the Sherman Act could be used to dissolve a monopolist.[14] A number of companies requested the attorney general to take action against Bell under the Sherman Act. Three types of complaints were made: (1) the complaint of Postal Telegraph that it had been discriminated against through the Bell–Western Union alliance because telegraph messages that came over the telephone lines were sent through Western Union, (2) the complaint of some independents that they could not get access to the long-distance lines, and (3) the complaint of some independents that the merger policy itself was illegal.

Rather than risk legal action that could be adverse to the system, the Bell system entered into negotiation with the attorney general and in December 1913 reached an agreement known as the Kingsbury Commitment. The specific event leading to the discussions was the plan to consolidate the Morgan purchased telephone companies in Ohio with the Bell system. The attorney general expressed an opinion that such a consolidation would be a violation of the Sherman Act and began discussions with Vail and Vice-President Kingsbury on a possible settlement. The Kingsbury Commitment was a unilateral letter rather than an actual consent decree. In the letter, AT&T promised to dispose of its Western Union stock, allow interconnection with the independents, and refrain from acquiring any more directly competing companies. The commitment removed the legal pressure from AT&T and improved relationships between AT&T and the independents. It did have direct costs to the company, both in direct dollar costs and in reduced freedom to use its market power. AT&T disposed of its Western Union stock three months after the Kingsbury Commitment at a loss of approximately $7.5 million (relative to the original purchase price) and also paid a fee of approximately $2.5 million to J. P. Morgan and Company as compensation for merger efforts that were abandoned as a result of the commitment.[15]

The Kingsbury Commitment prevented the complete takeover of the industry by Bell, but also reduced the competition between Bell and the independents. With interconnection, the independents gained an interest in the Bell system because the extensiveness of the Bell lines affected the communications of independent subscribers. The change in policy also made the independents less concerned about which properties Bell purchased because the purchases would not have a negative effect on them. The commitment was interpreted to allow the purchase of competing property so long as an equal number of stations was sold to an independent. This provision allowed Bell and the independents to exchange telephones in order to give each other geographical monopolies. So long as only one company served a given geographical area there was little reason to expect price competition to take place. Interconnection reduced Bell's ability to drive the independents out of business but also eliminated the independents' incentive to establish a competitive long-distance system. The Kingsbury Commitment was a compromise for the Bell system. It imposed limitations on the company's ability to directly control all of the telephones, but cost little in terms of profitability.

The Kingsbury Commitment also indicated Bell's recognition of a major change in political environment and public policy which had occurred since the original expiration of patents. A firm could not use any method it chose to strengthen market power but had to be sensitive to actual legal interpretations of the time and to public opinion that might lead to new laws or court decisions. Bell's behavior in the years just preceding 1913 was less aggressive than in the earlier years, but caused the company greater legal difficulties because of the changed political environment. Thus the commitment was an effort to exchange the short-run goal of full control of the telephone industry for an increased probability of retaining its existing industry dominance. In this case, the antitrust authorities favored competitors over competition because local monopolies were allowed so long as the independents were satisfied with them. The government served as a referee to narrow the gap in bargaining power rather than preserving true competition in the telephone industry.

The mergers continued after 1913 at a reduced rate. In 1912, the last complete year before the agreement, Bell purchased 136,000 telephones and sold 43,000. In the first three years after the agreement, Bell purchased an average of 30,000 telephones per year and sold an average of 11,000 per year. The mergers and trades increased during World War I while the government controlled the telephone systems as a war measure. However, many independents felt that the Kingsbury Commitment restrictions unduly limited their opportunities for a profitable merger. Consequently, Bell and the independents supported passage of the Willis-Graham Act of 1921, which exempted telephone mergers from antitrust review if approved by the regulatory authorities. The Willis-Graham Act ended the Kingsbury Commitment and induced another increase in the rate of telephone mergers. Bell's acquisitions hit a one-year high of 207,540 telephones purchased and 4,976 sold during 1927.[16]

Although antitrust enforcement prevented the completion of Bell's merger plan, the mergers succeeded in transforming the telephone industry from one evenly divided between Bell and the independents in 1907, with extensive direct competition and little cooperation, into a cooperative interconnected industry dominated by Bell with no direct competition. The sales to Bell and slow growth in remaining independent areas left the combined independents with approximately the same number of telephones in 1932 as they had in 1912, while the number of Bell telephones increased by 171 percent during that period. In the first twenty years after the Kingsbury Commitment, Bell purchased 1.7 million telephones and sold 0.4 million telephones for a net gain of 1.3 million. If all of the sales had taken place in 1912, they would have increased Bell's market share in that year from 58 to 73 percent. Bell's market share reached 79 percent in 1932 as a result of the mergers and more rapid growth than the independents.[17] The development of geographical monopolies and interconnection was more significant in reducing competition than actual market share changes. So long as there was only one long-distance network with all local companies connected to it, and so long as each local company had a geographical monopoly, the system could be operated in a unified manner as if it were all under one company's control even if no company had a dominant share of the nationwide total of installed telephones.

The sale of Western Union had a minor effect on AT&T's competitive strength. At the time of the Western Union purchase, long-distance telephone service was still restricted to approximately half the distance across the country and no effective telephone amplifier was in sight. The telephone system had a broader scope (exchanges in 50,000 cities and towns compared with Western Union offices in 21,000 cities and towns in 1909), but the technological restriction on the length of long-distance calls made people dependent on the telegraph system for coast-to-coast service.[18] Telegrams were also cheaper than long-distance calls. About the time Western Union was divested, the vacuum-tube-based amplifier became practical and coast-to-coast long-distance service was inaugurated. Further technological developments brought down the cost of long-distance service and relegated the telegraph companies to a secondary place in the communications system. While the divestiture of Western Union imposed some costs on AT&T, future technological developments proved the wisdom of the decision to give it up voluntarily rather than risk litigation that might have reduced AT&T's control of telephones.

The Emergence of Regulation

The second prong of Vail's strategy was to embrace regulation rather than to fight it. Five states had established some sort of regulatory control over telephone companies prior to 1907 (Mississippi, Louisiana, North Carolina, Virginia, and South Carolina), but in all five the regulation was nominal with no real authority over rates. Railroads had been regulated by the Interstate Commerce Commission (ICC) for almost twenty years. Public support for increased regulation was strong and seven new state regulatory commissions were established in 1907 alone.[19] The increasing political support for regulation and the early antitrust cases suggested that a private unregulated monopoly might not be allowed to exist. Thus the company could be forced to choose between loss of market control and regulation.

Regulation could provide positive benefits to the company. The regulatory agencies of the time were not very effective in controlling company behavior, and weak regulation could provide a justification for unified control of the system. Early pronouncements from some courts and regulatory agencies had indicated displeasure with competition in telephones because of the problem of running double wires along the streets and of lack of interconnection of the systems. Thus there was reason to believe that a regulatory agency would sanction the combination of Bell and its competitors and also prevent other companies from entering the industry. Enough experience had been accumulated through the Interstate Commerce Commission regulation of the railroads to see that regulation would not necessarily reduce profits.[20]

Vail began advocating regulation in the 1907 annual report in which he stated:

> It is contended that if there is to be no competition, there should be public control.
>
> It is not believed that there is any serious objection to such control, provided it is independent, intelligent, considerate, thorough and just, recognizing, as does the Interstate Commerce Commission in its report recently issued, that capital is entitled to its fair return, and good management or enterprise to its reward.[21]

In the 1910 report, Vail again discussed the benefits of regulation and asserted that a regulated monopoly could provide better and cheaper service than either competition or government ownership. He also explicitly discussed the need for protection from other competition: "If there is to be state control and regulation, there should also be state protection—protection to a corporation striving to serve the whole community . . . from aggressive competition which covers only that part which is profitable."[22] Vail echoed the same theme in a 1915 speech in which he said, "I am not only a strong advocate for control and regulation but I think I am one of the first corporation managers to advocate it. It is as necessary for the protection of corporations from each other as for protection to, or from, the public."[23]

The desire for regulation was not confined to the Bell system. When the extension of ICC authority to the telephone industry was being considered in 1910, the secretary of the National Independent Telephone Association wrote to a senator: "We do not ask the Government to fight our battles, but we do ask for protection against outrageous methods of warfare which are illegal and detrimental to the public welfare . . . We are not afraid of supervision; we believe in regulation."[24] With the support of both Bell and the independents, the Interstate Commerce Commission Act was amended in 1910 to bring interstate telephone companies under the jurisdiction of the ICC. The regulation of interstate telephone lines by the ICC was nominal and was confined to investigating complaints and prescribing accounting systems. The commission did not investigate the reasonableness of long-distance rates or order any changes in the Bell rates. Most states established telephone regulation in the years after 1907.

The combination of state and federal regulation stabilized the industry and ended the rate wars that had occurred during the early period of competition. Regulation increased the difficulty of new entry. The existence of regulation calmed public criticism of Bell's monopoly power and reduced the possibility of antitrust action against the firm. The regulation was quite variable during the early years, with no clear basis for setting rates. The general principle of fair return on assets was accepted but the method of computing the rate base was not. Consequently a variety of rates could be valid under the rules of the regulatory commissions. It appears that the regulation maintained Bell's existing profit rates at the time regulation was imposed. After a careful financial analysis of the industry, Stehman concluded:

> For the system as a whole, from 1900 on through the period when the various States, one by one, gave their commissions regulatory power over telephone companies, the earnings continued at about the same rate . . .
>
> . . . It can be said, however, that regulation, coming late and still almost non-existent in several of the States, has had relatively little effect in influencing the growth and financial success of the Bell telephone system.[25]

Effective regulatory control was impeded by three factors: (1) uncertain laws and constitutional authority with resulting appeals of adverse decisions to the courts, (2) the relationships between the operating companies and the parent company, and (3) the ownership of Western Electric. The laws giving regulatory authority were of course subject to the constitutional restrictions on government confiscation of property. This provided grounds for appeals through the court system of many regulatory decisions. The courts failed to articulate clear and precise rules to provide standards for future regulation and allowed endless controversies over the rate base and the fair return on capital. Thus the commissions could not take clear and decisive action and be certain of being upheld.

The relationship between the operating companies and the parent company provided additional problems. Many operating companies conducted business in more than one state, causing difficulty in determining the financial results of their business on a state-by-state basis. The parent company owned the operating companies and rented the telephone instruments to them. The telephone instruments were made by Western Electric, also a subsidiary of the parent company. The parent company was regulated at the federal level, the operating companies were regulated at the state level, and Western Electric was not regulated at all. The relationships among the companies made it effectively impossible for the regulatory commissions to determine the appropriate prices to provide a fair return on capital.

The inability of the state commissions to evaluate the costs of the operating companies and the nominal federal regulation through the ICC caused the primary impact of regulation to be slowing the rate of price changes rather than relating prices to the cost of service. The requirement to go through a hearing prevented rapid price changes, and was a real restriction when costs were rising rapidly as in the post–World

War I inflation. When costs were declining or steady so that there was no need for the company to raise its price, regulation had relatively little price effect. Declining costs were particularly characteristic of the long-distance service because of technological progress, and thus it was easy for the ICC to largely ignore the telephone industry.

By accepting regulation voluntarily, Bell reduced the risk that unfavorable regulation would be imposed. The system of competing federal and state regulation, together with the complex Bell structure, prevented real regulatory control while providing the protection and legitimacy of a regulated utility. Vail was careful to distinguish between regulation and management and was happy to accept regulation so long as it did not encroach on what were considered management prerogatives. The acceptance of regulation was a risk-reducing decision. It substituted a limited but guaranteed return on capital and management freedom for the uncertainty of the marketplace. It gave the Bell system a powerful weapon to exclude competitors and justification for seeking a monopoly, as well as reducing the chances of outright nationalization or serious antitrust action.

The Radio Challenge

The third leg of the AT&T strategy to restore its market power was an increased effort to develop protective patents. This was hardly a new strategy, as patents had played a fundamental role in the company development from the beginning. However, the period following 1907 witnessed a broader scope of scientific work than previously. The previous policy was to concentrate internal effort on engineering improvements and to systematically review developments outside the company for possible purchase. This had been a successful approach and had secured for the Bell company many key patents that were developed outside, including the Pupin loading patent. However, the strategy was risky because of the possibility of failing to secure patent rights to a major innovation that could displace telephones as telephones had displaced telegraph. If a major innovation should be developed, the company might be unable to purchase it or might be unable to recognize its significance if it lacked adequate research personnel of its own.

The increased emphasis on fundamental scientific research was not merely protection against abstract dangers from future developments, but was concerned with the potential competition posed by the development of radio. Guglielmo Marconi had patented his device for communicating over a distance without wires in 1896. Despite its undeveloped state, the radio was immediately recognized as a significant invention. In 1897 a group of wealthy Englishmen formed the Wireless Telegraph and Signal Company and paid Marconi £15,000 cash plus £60,000 in stock for his patent. The early Marconi radio put out a burst of waves of various frequencies. It could not be tuned to a single frequency and the waves could not carry voice signals. The device could carry telegraph codes and provided a method of communicating between ships and the shore. The Wireless Company carried on extensive experiments to improve the original device and successfully established communications between ships of the British Navy in 1899. In that year the Marconi Wireless Telegraph Company of America was formed as a subsidiary of the British company to carry out communications in the United States. The original Marconi patent, together with patents on improvements, gave the company a monopoly on wireless communication in the United States and Britain. By 1907 the radio was in common use on ships and was achieving some competition with underwater cables for overseas communication. It thus posed a direct competitive threat to the telegraph companies and a potential threat to the telephone.[26]

A related technological threat was the invention of the three-element vacuum tube, protected by 1907 and 1908 patents issued to Lee DeForest. The DeForest vacuum tube provided the key to improvements in radio as well as in long-distance telephone conversation, because it allowed efficient and accurate amplification of currents. The tube also allowed the transformation of radio technology from the Marconi "arc transmitter," which was useful only for telegraph, to a symmetric wave transmitter, which could carry voice as well. In 1907 DeForest organized the DeForest Radio Telephone Company to develop wireless telephone service. He carried out successful experiments and established a limited radio telephone service between ships of the U.S. Navy. DeForest's company later failed after a successful patent infringement suit was brought by the Marconi company.[27]

The opportunity and threat posed by the invention of radio and the vacuum tube caused AT&T to dramatically expand the quantity and quality of its research force. In 1909 the research director presented the following argument for additional capability:

> Whoever can supply and control the necessary telephone repeater will exert a dominating influence in the art of wireless telephony when it is developed. The lack of such a repeater for the art of wireless telephony and the number of able people at work upon that art create a situation which may result in some of these outsiders developing a telephone repeater before we have obtained one ourselves, unless we adopt vigorous measures from now on. A successful telephone repeater, therefore, would not only react most favorably upon our service where wires are used, but might put us in a position of control with respect to the art of wireless telephony should it turn out to be a factor of importance.[28]

Western Electric's research and development effort was expanded from 192 engineers in 1910 to 959 in 1915, while the research department of the parent company was also multiplied several times over. The increase in numbers was associated with an increase in scientific understanding and training of the personnel employed. The 1906 research report stated that "every effort in the department is being exerted toward perfecting the engineering methods; no one is employed who, as an inventor, is capable of originating new apparatus of novel design." In contrast, the 1911 research report stated that "it was decided to organize a branch of the engineering department which should include in its personnel the best talent available and in its equipment the best facilities possible for the highest grade research laboratory work . . . A number of highly trained and experienced physicists have been employed, beside a group of assistants of lesser attainments and experience."[29]

The expanded research force gave AT&T the ability to develop important technology for telephone operations, and also to develop patents on technology the company did not intend to use. A large portfolio of unused patents would allow the company to restrict the activities of companies in other fields and provide an opportunity for profitable trading of patent rights to create a better protected position. A policy was established of seeking patents on all of the alternative methods of accomplishing a technological objective, rather than only the method which seemed most useful. The chief engineer of Western Electric described the policy as follows:

> It was decided that we should adopt the policy of developing alternate methods to the more important lines of development, and carry

> on these alternatives to a point where patents might be secured . . . The plan is to keep the actual physical tests down to the point where simply enough information is gained to justify the prosecution of a patent application.[30]

The benefits the company could expect from acquiring patents on technology it did not expect to use were later described by an AT&T executive as follows:

> The regulation of the relationship between two such large interests as the American Telephone & Telegraph Co. and the General Electric Co. and the prevention of invasion of their respective fields is accomplished by mutual adjustment within "no man's land" where the offensive of the parties as related to these competitive activities is recognized as a natural defense against invasion of the major fields. Licenses, rights, opportunities, and privileges in connection with these competitive activities are traded off against each other and interchanged in such manner as to create a proper balance and satisfactory relationship between the parties in the major fields.[31]

AT&T's research effort convinced the company of the fundamental importance of DeForest's patent. Consequently all patent rights were purchased from DeForest, except for a personal right to use the device in his own enterprises, for a reported $250,000 in 1913. DeForest was in a relatively weak bargaining position given the importance of the vacuum tube. The most obvious uses for the tube were in telephone repeaters (amplifiers) or in radio, both of which were thoroughly occupied fields. In addition, the device required extensive development work to increase its power and regularity of operation. The Marconi company also had rights to the "Fleming Valve," a two-element vacuum tube that could be used for radio detection. The Fleming patent contained a broad claim for use of vacuum tubes in radio signal detection. The Marconi company sued DeForest in 1915 for infringing the Fleming diode and DeForest responded with a suit against Marconi for infringing his patent. The court ruled in 1916 that the DeForest tube was an infringement on the Fleming patent if used in radio and also that Marconi's use of a three-element tube was an infringement of the DeForest patent, thus preventing either company from using a three-element tube for radio.[32] Although the litigation was not completed at the time of the DeForest sale to AT&T, the prospect of litigation and further development work severely limited DeForest's ability to commercialize the invention on his own.

AT&T's purchase of the DeForest rights did not immediately give it the ability to make free use of the device. It was restricted in radio usage by the Marconi patents. The company found that the major improvement needed to make the vacuum tube usable in a telephone amplifier was a better vacuum than used by DeForest. AT&T personnel developed a high-vacuum version of the DeForest tube, but were then in interference with General Electric's patent application for high vacuum. General Electric also possessed other radio-related patents including the Alexanderson Alternator, which was the first practical device for symmetric, high-powered waves. Another limitation on AT&T's ability to make free use of the vacuum tube was Westinghouse's ownership of the Armstrong feed-back circuit patents, which allowed the production of high-frequency waves with a vacuum tube.[33] The complex patent situation prevented any one company from producing the best possible radio apparatus without licenses from other patent holders and created an ideal situation for patent negotiations to divide the field among the competitive parties.

The entry of the United States into World War I in 1917 greatly expanded the demand for radio. Radio was used for communications with merchant and navy ships and as an alternative to the Atlantic cables which were subject to being cut. The first regular transatlantic radio-telegraph service had been inaugurated in 1913 by the Marconi company. After the cables linking Germany and Britain were cut at the outbreak of the European war in 1914, the United States had been dependent on radio for communications with Germany. The Marconi company had been successful in preventing significant entry into the radio business through its basic patents despite the major technical advances made by others. In 1917 it still manufactured 95 percent of the radio apparatus.

Faced with the military need for radio and a difficult patent situation, the navy assumed responsibility for all patent infringement during the war and ordered all companies with the technical capability and desire to produce radio apparatus. Extensive progress was made in radio technology and manufacturing techniques during the war. Radio apparatus was manufactured by the Marconi company, General Electric (GE), Westinghouse, Western Electric, and others. The navy assumed operating control of radio service. The navy built new coastal stations for ship-to-shore communication and large high-power transmission stations for improved transatlantic service. The largest of these was a 1000-kilowatt transmitter set up in France with eight 820-foot masts supporting a mile-long antenna, which was not completed until after the armistice.[34]

At the end of the war, the navy supported a bill in Congress to leave permanent operating control of coastal radio stations to the navy, but it was defeated. The armistice also ended the navy's assumption of patent infringement responsibility and returned the companies to the prewar patent deadlock. The British Marconi company began negotiations to secure the patent rights necessary to build and use the best apparatus. High-ranking naval officers, concerned about British control of a military necessity, intervened in the negotiations and suggested that GE purchase the American Marconi company rather than sell the Alexanderson rights. Because the navy was the largest customer for radio apparatus, the suggestion had considerable weight even without force of law or direct navy financial intervention in the company. A company combining Marconi and GE capabilities would clearly be more valuable with the navy's good will and purchasing power than without it.

General Electric formed the Radio Corporation of America (RCA) in October 1919 as its radio operating subsidiary and purchased the assets of the American Marconi company on behalf of RCA. Following up the navy's concern about foreign control of radio, the initial RCA by-laws provided that at least 80 percent of the stock of the company would be held by American citizens or U.S.-controlled corporations. GE granted its patent rights in radio to RCA with the restriction that GE would be the exclusive manufacturing agent for RCA and RCA would be the exclusive purchaser of GE radio apparatus.[5]

The next step was to secure an agreement with AT&T for use of the now-indispensable vacuum tube. In July 1920 a patent cross-licensing agreement was signed between GE–RCA and AT&T–Western Electric. The patents were used to clarify the boundaries of the industries and licenses were granted for particular uses rather than for particular patents. AT&T received exclusive licenses for wire telegraph systems and wire telephone systems except for General Electric's right to establish private systems for itself and electric utilities. GE received an exclusive license for radio-telegraph operations (the then dominant market for radio). AT&T was given exclusive rights to develop radio-telephones as a public network while GE received exclusive rights to develop radio-telephones for operation of electric utilities and communication among ships, airplanes, and automobiles. Nonexclusive licenses were exchanged for transatlantic radiotelephony. At the same time, AT&T bought $2.5 million of RCA stock to give it a minority interest in the radio company. The following year the agreement was extended to include Westinghouse, which owned important circuit patents. In exchange for use of the Westinghouse patents, it was agreed that RCA would purchase 60 percent of its radio apparatus from GE and forty percent from Westinghouse. AT&T also received the rights to use Westinghouse patents in the activities reserved to it. The agreements covered some 1,200 patents.[36]

The patent exchange agreement benefited all parties. It gave General Electric the right to use the Marconi business it had purchased for RCA with outside patents and without fear of entry. It gave AT&T additional protection and rights on its wire telephone business and gave it the right to develop a radio-telephone business if that should prove economically feasible. AT&T thus was protected from outside entry into its main business through the new technology and given the right to exploit the new technology in its established business. The division of interests is similar to what would have happened if Western Union had succeeded in its 1879 proposal to the Bell company that Bell develop the local exchanges and Western Union the long-distance network. In that case, Bell would have had the right to develop the new technology in its then feasible form for local service, and Western Union would have been able to switch to telephone technology to protect its long-distance control when and if that became feasible. The probability of a radio-telephone long-distance network providing significant competition to the wire telephone long-distance network was as high from the 1920 perspective as the probability of a voice network providing substantial competition to a telegraph network from the 1879 perspective.

By dividing the patent rights along market lines, the agreement provided an effective barrier to entry into the respective markets. It produced a unified group of patents that would be difficult for an outsider to overcome. It reduced the probability that the patents would be invalidated by litigation among the parties. It gave all parties an interest in maintaining the entire group of patents rather than only the ones owned. AT&T saw its ability to trade its patent position for protection from radio competition as an important return on its research. Soon after the GE agreement, the AT&T chief engineer described the radio research as follows:

> It seems to me that this enlarged and enhanced position played no small part in enabling us to reach our present satisfactory understanding with the General Electric Co. and the Radio Corporation of America and that if we never derive any other benefit from our work than that which follows the safe-guarding of our wire interests we can look upon the time and money as having been returned to us many times over.[37]

The 1920 agreement had seemed to clarify the radio boundaries and provide clear lines between the competing interests, but it was drawn in terms of point-to-point communications and did not contemplate the development of radio broadcasting. The first broadcasts have been attributed to Frank Conrad of Westinghouse who played records at regular times over the transmitter that he was developing for the navy. A group of amateurs with home-made sets began listening for the Conrad tests. The resulting interest in radio sets caused Westinghouse to set up a radio station in Pittsburgh at the end of 1920 to stimulate purchases of radio equipment.[38] The experiment successfully increased the demand for receivers and during 1921, Westinghouse established stations in New York and Boston, while RCA and other companies began planning broadcasting stations of their own.

The development of broadcasting as the major radio use caused the original agreement to be ambiguously related to the market situation. AT&T had received exclusive rights to the development of public service telephony, while GE–RCA–Westinghouse had rights to radio telephone for internal uses and exclusive rights to produce equipment for amateur radio. Because those people who were listening were not commercial operators, they fell under the definition of amateurs and thus GE–RCA–Westinghouse appeared to have exclusive rights to produce receiving sets. The radio companies were also entitled to produce transmitters for their own use; but, because commercial transmitters would be classified as a public radio-telephone network, AT&T would have the rights to sell transmitters to broadcasters other than the GE–RCA–Westinghouse group.

Broadcasting spread rapidly during 1922 and RCA sales jumped from $1.5 million in 1921 to $11.3 million in 1922. AT&T decided that radio broadcasting was an adjunct to telephone service and should be controlled by the company. AT&T began to establish its own radio stations interconnected with telephone wires to form a network. Telephone wires for connection of radio stations or for remote pick-up and transmission to the studio were denied to competitors. Under the patent agreement, AT&T had a right to make receiving apparatus "for direct use in connection with transmitting apparatus made by it" in order to allow the company to develop two-way radio. The company consequently built sets with fixed reception frequencies tuned to AT&T broadcast stations to make them qualify under the patent rights as apparatus used directly with AT&T transmitting equipment.[39] The director of radio for AT&T stated the company's intentions as follows in 1923:

> We have been very careful, up to the present time, not to state to the public in any way, through the press or in any of our talks, the idea that the Bell System desires to monopolize broadcasting; but the fact remains that it is a telephone job, that we are telephone people, that we can do it better than anybody else, and it seems to me that the clear, logical conclusion that must be reached is that, sooner or later, in one form or another, we have got to do the job.[40]

The AT&T attempts to take charge of broadcasting were not welcomed by RCA. RCA claimed that AT&T was violating the patent exchange agreement and called for formal arbitration as provided for in the agreement. RCA challenged AT&T's right to refuse pick-up lines to transmit programs to RCA studios and AT&T's right to sell pretuned receivers. The referee ruled in RCA's favor in Novemer 1924. However, AT&T refused to accept the arbitrator's interpretation of the agreement because it increased the probability that the agreement violated the antitrust laws.

Rather than continue to dispute the provisions of the 1920 agreement, the parties negotiated a new agreement in 1926 that set out the fields open to each more clearly. The agreements were complex, but in essence AT&T agreed to abandon radio broadcasting and receiver sales to RCA in exchange for a strengthened position in point-to-point communication. Three contracts were signed on July 1, 1926. The first sold the Broadcasting Company (AT&T broadcasting operations) assets to RCA (where it became a part of National Broadcasting Company), and agreed to pay RCA $800,000 if AT&T reentered broadcasting before 1933. The second agreement provided that RCA would obtain program transmission wire service from AT&T, eliminating the possibility that RCA could use AT&T patented devices to set up its own network as an adjunct to radio broadcasting. So long as AT&T intended to control the broadcast stations, denial of connecting links was a useful strategy for the company because it increased AT&T's competitive edge in broadcasting. However, if RCA was to own and operate the network of broadcasting stations, forcing it to set up its own wire distribution network was only denying business to AT&T and causing the creation of a potential telephone competitor.

The third agreement modified the original cross-licensing agreement. Bell retained exclusive rights to wire telephone and telegraph service, and was given some ship-to-shore rights (previously reserved to RCA–GE) and exclusive rights to two-way radio communication (previously shared). Nonexclusive licenses were exchanged for sound motion pictures, leaving that field open to competition between the groups. Bell received exclusive rights for sound picture equipment connected with wire transmission, while the radio group received exclusive licenses for sound picture equipment connected with one-way radio. In other words, the agreement extended to the foreseen but undeveloped television technology and agreed to divide the field as with radio. RCA received the television broadcast rights and AT&T received the television wire transmission rights.[41]

In May 1930 the government filed antitrust charges against AT&T, GE, Westinghouse, and others alleging that the 1926 agreement was a conspiracy in restraint of trade in violation of Section 1 of the Sherman Act. An out-of-court settlement was reached in November 1932 in which the license agreement was changed from exclusive licenses to nonexclusive licenses. Subsequently a new license agreement was drawn up between the parties prohibiting them from enforcing the exclusive provisions of the previous cross-licensing agreement. The consent decree also required GE and Westinghouse to dispose of their RCA stock.[42] AT&T had previously sold its stock in RCA. The change to nonexclusive licenses does not appear to have had any significant effect on the operation of the agreement. The president of Bell Laboratories described the 1932 agreement as follows:

> Broadly speaking, the practical effect of the agreement is to limit the field of possible development of each party to its present major activities. Where unilateral grants only are made it is clearly impossible for the granting party to contemplate successful competition when confronted with an adversary who has both its own and the other per-

> son's patents. Even where bilateral licenses are made, there is probably little danger of competition by the grantee in fields where the grantor has already attained to a commanding position. Thus, while a casual reading of the agreement by one not thoroughly conversant with all the factors may appear to establish the basis for an enlarged free development in most of the fields, this is not actually the case.[43]

The 1926 agreement gave both RCA and AT&T rights under the combined patents to develop sound motion picture equipment. Sound pictures were recognized as a potentially profitable market but did not fall naturally into the province of either company. AT&T decided to make an initial effort to dominate the motion picture field. The head of the company's motion picture activities in 1927 wrote:

> Competition . . . will doubtless ultimately result in a situation highly favorable to the motion picture interests and opposed to our own. This is an extensive and highly profitable field and it is quite worth our while to go a long way toward making it practically an exclusive field. I believe that we could justify, from a commercial standpoint, paying a large price for the liquidation of the Radio Corporation for this purpose alone.[44]

No action was taken on "the liquidation of the Radio Corporation" but AT&T (through its subsidiary ERPI) did take a number of actions to secure a controlling interest in motion pictures before bowing out of the field under antitrust pressure. In 1928 the major motion picture producers (who had agreed among themselves to bargain as a group for sound equipment) chose ERPI production equipment over that offered by RCA. The producers also agreed to install ERPI exhibition equipment in the theaters owned by them. Restrictions were written into contracts for ERPI production equipment prohibiting the exhibition of films produced by that equipment on exhibition equipment other than that manufactured by ERPI. Similarly, users of ERPI exhibition equipment were prohibited from showing films not produced on ERPI equipment. A later relaxation of the requirement to use ERPI or equal quality equipment had little effect because of ERPI's refusal to certify RCA equipment as equal quality. Distributors who were licensed to use ERPI equipment were required to pay a royalty to ERPI if they distributed films produced on RCA equipment.[45]

The restrictions transformed an ordinary industrial market into one with systems characteristics similar to those of the telephone market. So long as the dominant film makers used ERPI equipment, the theater owners had an incentive to acquire ERPI exhibition equipment in order to maximize their access to the films available. Similarly, so long as the dominant exhibition equipment was ERPI, the film makers had an incentive to use ERPI equipment in order to maximize their potential viewing audience. Thus, if the system could be established, it would be difficult for any competitor to get an order for either production or exhibition equipment. On the other hand, if the ERPI equipment became minority owned, the restrictions would be counter productive and would encourage users to drop it in favor of the majority owned equipment. Thus the restrictions were an attempt to strengthen the position gained by the initial order for equipment from the major producers.

The restrictions evoked a storm of controversy from RCA and those who preferred to use RCA equipment. Toward the end of 1928, RCA engaged a law firm which rendered an opinion that the restrictions were illegal and threatened to file an antitrust suit. In addition, one of the theater companies filed an antitrust suit against producers licensed by ERPI for refusal to supply films. Following the legal presure, ERPI dropped the restrictions on interchangeability in December 1928. ERPI's share of the motion picture exhibition market dropped from 92 percent at the end of 1928 to 40 percent of the substantially larger market in 1929. While the abandonment of the restrictive viewing clause eliminated the most serious complaint with ERPI policy, the requirement that ERPI equipment users obtain all service and parts from ERPI and the "double royalty" provision requiring ERPI-licensed distributors to pay ERPI royalties on the distribution of films produced on RCA equipment caused extensive litigation. A total of twenty-two antitrust suits were filed over the restrictions by 1936. ERPI eliminated the double royalty provision in 1935 and dropped the repair and replacement restrictions two years later in compliance with a court order. Following the end of the restrictive policies, ERPI granted licenses to other companies to produce the motion picture apparatus under its patents, sold its service business, and abandoned active participation in the motion picture industry.[46]

AT&T's patent strength and direct participation in peripheral areas, such as radio broadcasting and motion picture equipment, provided it with bargaining chips to defend its primary area of telephone service. So long as it had strength in areas of interest to potential telephone competitors, it could trade an explicit promise to abandon an area of peripheral interest for a promise from the potential competitor to stay out of the telephone business. An AT&T executive described the strategy as follows in 1927:

> A primary purpose of the American Telephone & Telegraph Co. is the defense and maintenance of its position in the telephone field in the United States. Undertakings and policies must be made to conform to the accomplishment of this purpose.
>
> The American Telephone & Telegraph Co. is surrounded by potentially competitive interests which may in some manner or degree intrude upon the telephone field . . .
>
> If the American Telephone & Telegraph Co. abandons its activity in the commercial competitive field and other potentially competitive interests continue their activities, it means that they will carry their offensive right up to the wall of our defense and our trading must be in our major field against activities in their outlying commercial fields. The nearer the trading can be carried to the major field of our competitors the more advantageous trading position we are in.[47]

The research and patent strategy was successful in restoring the patent walls around AT&T that had worn down during the competitive period. AT&T's increased internal technical expertise and willingness to purchase key outside patents protected it from competitive attack through advanced technology. Its broadranging research activities gave the company an interest in a number of fields that it could trade to potential competitors for additional protection of the telephone field. The cross-licensing agreements of 1920 and 1926 significantly increased the patent barrier to entry by creating a huge pool of patents under which the companies were licensed to perform specific activities without regard for actual ownership of the patents. This removed the incentive for the companies to challenge each others patents and increased the incentive for them to assert a broad construction on the patents against any outside entrants. By 1935 the Bell system owned 9,255 patents and was licensed under 6,000 patents owned by others. The huge number of patents made it practically impossible for an entrant to be certain it was not infringing an AT&T patent. AT&T's annual research expenditures increased to a high of $22.6 million in 1930 before dropping again during the 1930s.[48] The ability to charge research expenditures to operating companies as an expense for rate regulation purposes while building barriers to new competitors made the research and patents strategy an efficient way to increase market power.

Evaluation of Market Power Restoration

The year 1907 was the high point of competition in the telephone industry. In that year the independent phone companies had almost half of the installed phones in the country. Western Union was a strong company with an extensive network of wires that formed a potential entrant into long-distance connecting service as well as providing telegraph competition. The increasing development of radio in companies outside of the Bell system posed an immediate competitive threat to telegraph and a potential competitive threat to telephone service. Radio created the possibility of establishing a rudimentary long-distance network to connect competitive telephone companies without extensive investments in wires or rights of way. The Bell system's narrowly focused research efforts had done much to improve telephone service but little to protect the company from invasion of new technologies. The competitive price policies had reduced profits for Bell and independent alike, but showed no signs of driving the independents out of business.

To restore market power in telephones, AT&T had to eliminate the direct competition of other telephone companies and increase barriers to entry. Mergers were the easiest method of eliminating direct competition, but they would have been of little use if they induced further entry or antitrust action. Thus an adjunct to the merger policy was the acceptance of regulation. State regulation allowed the mergers to proceed without antitrust attack. The exchange of telephones to give each company a geographical monopoly when mergers did not occur also eliminated direct price competition. It is unlikely that that program could have survived antitrust attack in the absence of regulation. Regulatory prohibitions on entry prevented the merger policy from attracting new entrants who hoped for favorable merger terms.

The combination of mergers, customer exchanges, and state regulation eliminated direct price competition. AT&T's cost of attaining market power through regulation was a limitation on the rate of price and service changes. The early experience with regulation was consistent with the protection of property rights in the status quo as developed in Chapter 1. Uncertain authority, inadequate information, and the complex AT&T structure prevented the regulatory authorities from prescribing telephone rates based on costs even if they wanted to.

The regulators could and did take the existing rates as given and require justification for changes in those rates. The regulatory process slowed all changes and gave interested parties a forum to negotiate changes rather than having them imposed unilaterally by the company. In times of inflation, the regulatory process held rates down because rapid changes were not allowed. In times of stable or decreasing costs, the regulation probably had little effect on the rates charged.

While the combination of mergers and regulation protected AT&T from existing competitors using the existing technology, it did not protect the company against major technological change. Technological change could threaten either by allowing another company to enter existing types of telephone service or by creating a new product that would take business away from the telephone. Radio and the vacuum tube threatened both kinds of change. Because the vacuum tube was crucial to a telephone amplifier for long-distance service, a company that controlled the vacuum tube could provide severe long-disance competition. Because radio could communicate without wires, it could lead to a new product to displace wire-based telephone systems.

AT&T's increased internal research sophistication and purchase of the DeForest vacuum-tube patents allowed the company to maintain its position despite changing technology. Without the increased internal research capability, AT&T likely would have missed the significance of the DeForest patents. At the time of the purchase, the tube was unreliable and unsuitable for telephone amplifier use. A very applied engineering approach would have rejected the DeForest tubes as inappropriate components for telephone systems. An appreciation of the extraordinary importance of vacuum tubes for telephone, radio, and other electronic devices required the ability to predict methods to improve and utilize the tubes. AT&T's early recognition of the vacuum tube's significance and its purchase of the patent rights provided the company access to the technology of potential competitors who needed vacuum tubes along with explicit promises that they would not enter telephones. Consequently AT&T was protected from technological displacement, not only by the vacuum tube itself, but by many other improvements covered by the patent pool.

Although AT&T significantly strengthened its market power after 1907, it did not accomplish all that it set out to do. The company failed to acquire all the independent telephone companies and complete its telephone monopoly. It failed to monopolize radio broadcasting and abandoned the field. It failed to monopolize sound motion picture production. The failures were directly related to the successful maintenance and extension of power over the telephone industry. Abandoning some goals reduced the risk that telephone control would be lost. The Kingsbury Commitment, which reduced the acquisition of independent companies, lessened the risk of a major antitrust suit directed at AT&T's control of telephones. Giving up broadcasting prevented the dissolution of the patent exchange agreement with resulting risk of entry into telephone competition. The exit from sound motion pictures reduced antitrust pressure on the firm. Entry into peripheral areas followed by judicious retreat allowed AT&T to develop bargaining chips without risking the loss of its telephone business.

Antitrust came into play several different times; the prohibition on AT&T's purchase of a competitive telephone equipment supplier, the restrictions on mergers, the requirement to connect the independents with the long-distance network, the elimination of the exclusive license provision of the 1926 patent exchange agreement, and the elimination of restrictive practices on movie equipment. It seems likely that without antitrust action, AT&T would have eventually taken over all of the independent telephone companies. Each merger increased AT&T's bargaining power relative to the remaining companies because of the increased number of phones that could be reached through merger and the greater disadvantage of being an isolated system. The interconnect requirement reduced the incentive to merge and probably had a greater effect than the merger restrictions as such. It is unclear whether or not AT&T would have been able to maintain the restrictive clauses in its movie equipment absent RCA's ability to use antitrust as a threat because of the rapid growth and technological change in that industry at the time the attempt was made. However, if the company had become established on the basis of restrictions between mixing production and exhibition equipment of different manufacturers, it would have presented a very difficult barrier to entry for new competitors and might have become a major business for AT&T.

7 The Era of Regulated Monopoly, 1934–1956

In 1934 AT&T was thoroughly established as the controlling firm in the telecommunications industry. It owned approximately 80 percent of the telephones in the United States and the only significant long-distance telephone network. The remaining 20 percent of the phones were owned by a large number of scattered small telephone companies that were dependent on the AT&T network for long-distance service and connections with other companies. Local telephone companies operated as geographical monopolies under state regulatory control. Barriers to entry were high in both local service and long-distance service. The local service barriers to entry were primarily determined by regulatory prohibitions on entry, with secondary protection coming from AT&T's patents and the systems advantages of an established firm. The long-distance barriers were primarily based on AT&T's patent control of long-distance technology through its own patents and its patent licensing agreements with other companies. Secondary protection came from the company's control of rights of way to lay wires and its systems advantages.

In 1956 the industry structure appeared to be practically identical to what it was in 1934. AT&T was still the controlling firm with approximately 80 percent of the telephones and the only long-distance network. Barriers to entry were still very high. However, the period witnessed a significant change in the basis of the company's market power. By 1956 AT&T's protection from competition at the national level was almost entirely due to federal regulation rather than patents and other market barriers. Federal regulation allowed the company to defeat two significant challenges to its dominance during the period: a major technological innovation that reduced the market barriers to entry and a federal antitrust suit that sought to dismember the company. AT&T voluntarily relinquished its patent protection in exchange for greater regulatory protection. Its activities during the 1934–1956 period transformed AT&T into a quasi-public company whose fortunes were more dependent on political and regulatory decisions than on market forces.

The Formation of the Federal Communications Commission

The initial political pressure for a revised communications regulatory structure came from a desire to centralize regulatory authority over communications. The Interstate Commerce Commission had jurisdiction over common carriers, while the Federal Radio Commission (set up in 1927 to control broadcasting interference) had authority over the assignment of frequencies and the executive branch had control over cable landing licenses. A bill to centralize the functions under one commission was introduced in 1929 and extensive hearings were held but no action was taken.

Political support for increased government control of the economy developed during the 1930s as more and more people lost confidence in the self-regulating aspect of a market economy. The communications industry was one of several selected for careful scrutiny and potential government action. Early in the Roosevelt administration, a committee was set up under the secretary of commerce to consider alternatives (including government ownership) to the existing communications industry structure. The committee recommended a system of strengthened and centralized regulation. President Roosevelt transmitted the report to Congress in February 1934. He asked Congress to establish a Federal Communications Commission with powers similar to those possessed by the existing separate agencies and authorization to conduct an investigation of the industry and make recommendations for further authority. Congress then passed the Communications Act of 1934 in June.

Although the law essentially followed Roosevelt's request to maintain existing practices in a centralized agency pending further study, several significant changes were made. Rate regulation authority was increased by requiring the carriers to post tariffs in advance and giving the commission authority to suspend tariffs announced by the carriers. The previous ICC authority to require advance approval of new transportation (but not communications) facilities and services was extended to FCC advance approval of communications facilities and services. The commission was also given the authority to require interconnection after a proper hearing and a determination that the action was "necessary or desirable in the public interest."[1] The power to compel interconnection, to suspend rates pending an investigation, to allocate frequencies, and to require prior approval for any expansion of facilities provided the framework for moving long-distance communications from a market-oriented industry to a politically oriented industry.

The 1934 law did not make any immediate difference in the operations of the telephone company. No criteria were set up for determining whether rates were "just and reasonable," and no general rate investigation was instituted. Instead the commission began a process known as "continuing surveillance" in which informal negotiations were carried on between the commission or its staff and representatives of the telephone company to determine the appropriate level of interstate rates. Soon after the commission was formed, AT&T made voluntary reductions in the interstate long-distance rates which amounted to $30 million per year at the existing volume of business. The commission took credit for the reductions.[2] However, there is some uncertainty over the relationship, if any, between the commission's activities and the rate reductions. The reductions were not ordered by the commission. They were made at a time of generally falling prices and technological progress and thus could have been a normal response to changing market conditions. The technological progress that had been continuous since the beginning of the telephone continued after the formation of the FCC and allowed the company to avoid requesting rate increases for long-distance service. The FCC was able to take credit for continually lower prices on long-distance service without undertaking any formal action to reduce the rates. No full-scale rate investigation was instituted until the 1960s.

The commission undertook an extensive investigation of the telephone industry and its history as part of its mandate to determine whether or not further legislation was necessary. The investigation cost the government $2 million and occupied three hundred researchers for several years. The staff developed an extensive history of telephone operations and AT&T practices. The "Proposed Report," filed in April 1938, advocated abolition of the close tie between Western Electric and the operating companies by requiring the operating companies to use competitive bidding for telephone supplies. It also recommended direct regulation for Western Electric. The proposed report estimated that Western could cut prices 37 percent and still earn an adequate return on its capital. AT&T launched an extensive campaign to stir up public support for the Western Electric operating system tie and the recommendation was deleted from the final report issued in June 1939. In the final report, the FCC requested several minor changes in the law but nothing that would change the character of regulation from that developed in the 1934 act.[3]

The Microwave Challenge

The simultaneous development of television and microwave transmission technology in the years immediately following World War II provided a significant threat to AT&T's control of long-distance communications. The advent of television created extensive new demand for high-capacity trunk channels to transmit television programs among television stations. A television program could not be carried on ordinary wires because it required the communications capacity of approximately one thousand telephone conversations. Television programs could be carried over a coaxial cable, developed by AT&T during the 1930s to carry a large number of telephone conversations simultaneously, or over a microwave relay system. Coaxial cable had many of the same characteristics as wires; it required a physical path over a right of way between two fixed points and was thus limited to companies that had or could obtain right-of-way privileges. Microwave relay was a radio-based system that required a relay tower at intervals of twenty to thirty miles and did not require a right of way between the points connected. The simultaneous availability of a significant new source of demand and a new technology that did not require rights of way provided the ability and the incentive for new firms to enter the long-distance telecommunications market.

Microwave radio systems were initially developed for the four to six GHz bands of electromagnetic radiation. Those bands are over three thousand times the frequency of AM radio and produce very short waves that travel in straight lines and can be focused. The ability to use microwave frequencies made long-distance telephone communication by radio practical forty years after the Bell system first began research on radio in 1907. However, by the end of the war, the initial radio patent pool no longer protected Bell from radio competition to its long-distance network. Microwave required not one single technical breakthrough but a variety of advances in technology in order to allow the practical utilization of higher and higher radio frequencies. Many companies contributed to its development. The Bell system, RCA, and others had done experimental work on microwave technology before World War II. The technology was advanced during the war as a result of radar developments, an application of microwaves. During the war, the Signal Corps used microwave radio for voice communications.

At the close of the war, microwave was known to be a viable technology for either television or multiple voice transmission but further development was required to produce a commercial system. Because much of the work had been done for the government, microwave technology was relatively free of patent control. The equal start of many different companies on microwave, the absence of patent protection, and the absence of right-of-way requirements eroded the barriers to entry for long-distance communication. However, microwave did require frequency allocations from the Federal Communications Commission. Because the frequencies were not being used before the development of microwave, they were freely available for the new technology, but the allocation procedure provided an opportunity for the established firm to seek protection from potential competitors.

Television technology was developed during the 1930s primarily by RCA. Zenith, AT&T, and others also carried out television-related research. A public demonstration of RCA television was given in 1939 and some receivers were sold to pick up a limited number of programs broadcast in New York. Disagreements over the proper standards for television delayed its final approval before the Federal Communications Commission until April 1941, but by that time the coming war prevented extensive commercialization. At the end of the war, there were nine part-time broadcasting stations reaching a total of 7,500 television sets, most of which were located in metropolitan New York.[4] The industry expanded rapidly following the war leading to a demand for circuits to transmit programs among the television stations.

Because existing communications networks were inadequate for television transmission, the television demand created an incentive for new entry into long-distance communications. The patent license agreements among AT&T, RCA, GE, and Westinghouse had foreseen a continuation of the radio arangement in which AT&T provided network services while RCA and others provided programming and receiving sets. However, the development of microwave together with antitrust modifications of the patent license agreements left the issue unsettled. In 1946 Zenith repudiated RCA licenses for the patent package that had formed the basis of radio industry organization and filed antitrust charges against RCA. This led to similar action by other radio licensees and extensive antitrust and patent infringement litigation that was finally settled in 1957 by a RCA defeat and payment of damages to Zenith. The cost of microwave relative to long-distance telephone costs also made private networks economically attractive to companies with high-volume internal communications.

Initially the FCC freely granted experimental licenses for the required frequencies, but deferred the question of permanent licenses and operating rights until the technology was more fully developed. This policy decreased the incentive of potential entrants to build networks by increasing the uncertainty associated with microwave technology, but the policy did not prevent several companies from participating in the experimental stage. Philco achieved the first operational domestic microwave link when it completed a Washington–Philadelphia line in the spring of 1945. Western Union followed soon afterward with a New York–Philadelphia link as the first step in a planned nationwide network. General Electric and IBM obtained licenses for a Schenectady–N.Y.–Philadelphia–Washington system in November 1944. The IBM–GE system was planned for business data transmission rather than television. AT&T opened its first microwave link (between New York and Boston) in November 1947. Raytheon planned a transcontinental system and obtained licenses for a Boston–N.Y.–Chicago system.[5]

AT&T had planned to increase telephone capacity and to develop television capacity with coaxial cable technology. After the successful transmission of television signals from Washington to Philadelphia over Philco's microwave system in 1945, AT&T placed microwave development on a crash basis and undertook an extensive research program to develop a technologically advanced microwave system. Some of the coaxial cable expansion that had been planned was deferred in favor of microwave.[6]

The early competition in microwave centered on two questions: (1) How would microwave frequencies be allocated by the Federal Communications Commission? and (2) Was AT&T required to provide service to all who requested it regardless of the origin of the programming? In the initial allocation of experimental microwave frequencies, AT&T's request that frequencies be allocated exclusively to common carriers was denied. AT&T continued to advocate the limitation of microwave frequencies to common carriers. By 1948 the commission adopted AT&T's position and reserved permanent use of the microwave frequencies to common carriers. At that time, AT&T still lacked facilities to meet the demand of television stations for interconnection. Consequently the television networks were authorized to build temporary microwave networks to meet their own needs, but only until common-carrier facilities became available.[7] The frequency allocation was a dominant event in the microwave competition because it ruled out permanent competition from companies not classified as common carriers by the FCC, while other events made common-carrier status undesirable or impossible for potential competitors.

Related to the license debate was a dispute over meaning of the experimental licenses. The original experimental licenses could have been limited strictly to experiments or simply treated as temporary licenses without restrictions on use during the experimental period. At the request of Western Union, the FCC adopted the first interpretation. It restricted the experimental licenses of IBM and GE to purely experimental activities and precluded them from using the channel commercially. The FCC interpretation prevented potential competitors from building a private network under an experimental license and using it to challenge the established carriers if a permanent license could be secured later.

In March 1948 AT&T filed its first video transmission tariff with the Federal Communications Commission. The tariff provided a basic service of eight consecutive hours per day, seven days per week of transmission. It also arranged an allocation scheme among potential users when a shortage of capacity made full service unavailable. The tariff included a strict noninterconnection policy. The policy prohibited not only direct physical interconnection of the facilities but also such things as receipt of a program in the studio over Bell lines and filming it from a television monitor for retransmission over private lines.[8] Any transmission that gave the impression of through transmission using both AT&T and non-AT&T facilities was defined as prohibited interconnection. An AT&T witness testified that if a program were delayed for a half hour in the receiving studio before retransmission, it would be allowable under the tariff. Interconnection was allowed when AT&T could not provide the facilities so long as the customer would switch back to AT&T as soon as the through facilities became available. The interconnection prohibitions were applied both to facilities owned by the television broadcasters (non-common carriers) and to those owned by Western Union (a common carrier).

In hearings on the tariff, the Bell system asserted that a monopoly in video transmission was socially desirable. An AT&T witness testified:

> Q. Just so that the record is perfectly clear on this point, then, it is the position of the Bell System that there should be a nationwide monopoly in one carrier of intercity television transmission facilities, and that one carrier should be the Bell System?
>
> A. Yes, I think the Bell System is in the best position to provide these nationwide television networks, and I think that will be in the best interest of the public to have Bell System provide it. I think they can do it economically and do a satisfactory job as indicated in experience with program facilities and private line facilities and nationwide message toll facilities.[9]

The witness continued with the statement that "competition with the Telephone Company in the intercity video field is not in the public interest."[10]

AT&T's interconnection restrictions were challenged by the Television Broadcasters Association. The FCC then began a hearing on the tariff. The hearing initially covered a wide range of issues but was later restricted to the interconnection issue in order to expedite the proceeding. While the hearing was under way, the tariff went into effect as scheduled. In September 1948, while the FCC hearing was under way, NBC received a telecast in New York from Philadelphia over the Philco system and delivered it to AT&T for transmission on to Boston. AT&T refused because of its tariff prohibition against interconnection. Similarly AT&T refused to deliver a program to NBC in New York when NBC planned to transmit it on to Philadelphia via the Philco facilities. Philco filed suit against AT&T and requested an injunction requiring AT&T to provide the services, but the court declared itself without jurisdiction until the FCC had acted. It stated that in order to have jurisdiction, "there must be some vice in the regulation other than unfairness, unreasonableness or discrimination" which were issues properly left to the commission to decide.[11] Subsequently, Philco announced that it was withdrawing from the microwave field as a result of interconnection refusal.

After a twenty-month FCC investigation of the legality of the AT&T interconnect prohibition, the FCC reached a decision in December 1949 that required interconnection with broadcast company systems but not with common carrier systems (Western Union). The theory behind the decision was that the 1948 permission for broadcasters to establish temporary microwave systems where the common carrier facilities were inadequate implied that those systems could be used in conjunction with common carrier facilities in order to make them meaningful additions to the common carrier capabilities. Thus it was a violation of previous FCC policy to deny connection between the broadcast company facilities (which the FCC reaffirmed were to be temporary and phased out as soon as adequate common carrier facilities were available) and common carrier facilities. However, with regard to Western Union, the commission emphasized the requirement in the Communications Act that the commission could order physical interconnection where that was found to be "necessary or desirable in the public interest" after a proper hearing, and ruled that the completed hearing was inadequate to establish a general need for interconnection among common carriers. Consequently the commission deferred resolution of interconnection between Western Union and AT&T for a further hearing.

In the 1949 decision, the commission stated:

> The Commission emphasizes that the policy . . . does not mean that intercity video relaying is to be restricted to a single common carrier. The Commission will consider applications by any qualified persons, including current operators of private facilities, for authority to operate microwave radio relay stations for the purpose of furnishing intercity video transmission service on a common carrier basis on the appropriate common carrier frequencies.[12]

However, the decision effectively did restrict the transmission to one carrier. By affirming the temporary nature of licenses not issued to common carriers, the commission forced potential competitors to become common carriers. By restricting the interconnection requirement to private systems, the commission required potential competitors to avoid common carrier status for short-run viability unless they could build a complete nationwide system. The option of building a complete system was of doubtful feasibility for a new entrant because of the enormous costs and the likelihood that the FCC would refuse authorization for competing common carrier facilities when the existing facilities of AT&T became adequate. None of the private video carriers became common carriers and only the Western Union system was left as potential long-term competition to AT&T.

Western Union stopped the expansion of its system while awaiting the outcome of the new hearing on common carrier interconnection. The second proceeding continued for three years until a decision was reached in October 1952. The commission placed a stringent interpretation on the language that interconnection could be required when in the public interest and ruled that, because there were adequate Bell facilities over the route actually served by Western Union, there would be no public interest in connecting Western Union into the network. The commission also ruled that the future routes of Western Union could not be considered because the requirement for a hearing implied specific facts and thus each route would only be determined on a case-by-case basis if the facilities were actually built.[13]

While the 1952 decision effectively eliminated the threat of competition for trunk-line video service, it did not totally stop the development of microwave systems outside of AT&T. Right-of-way companies (railroads and pipelines) built several systems for their own use without interconnection and without AT&T opposition. The development of Community Antenna Television Systems (CATV) to provide television service to places not served by regular broadcast television led to a demand for short microwave systems to connect an antenna with the wire distribution system. The first application was filed in 1952 and after three years the FCC granted an experimental one-year authorization. No interconnection was required, no direct competition occurred, and AT&T did not oppose the application. A number of short microwave systems in poor television reception areas were later established.

A third type of non-AT&T microwave system began with the 1954 request of two North Dakota television stations to operate a private microwave system where AT&T channels were available but at higher cost than the proposed system of the broadcasters. After four years of deliberation on the high costs of using regular tariffed service in areas remote from the trunk lines, the FCC amended its 1946 ruling restricting the use of private systems to places where no common carrier facilities were available. The 1958 rules allowed private broadcasting systems so long as no interconnection was required with the facilities of AT&T.[14] The prohibition on interconnection prevented any renewed challenge to AT&T dominance in the video relay market, but the ruling indicated that the frequencies were not so scarce as had been implied by the original restrictions and opened up the possibility of additional competition in later years.

The Antitrust Suit

The Justice Department filed a Sherman Act antitrust suit against AT&T and Western Electric on January 14, 1949, based on information gathered in the Federal Communications Commission investigation of the 1930s. The Justice Department lead attorney on the case had been one of the FCC principal investigators.[15] The seventy-three page complaint charged Western Electric and AT&T with a conspiracy to restrain trade in violation of Section 1 and charged Western Electric with monopolizing the market for telephones and related equipment in violation of Section 2 of the Sherman Act. The government asked for an end to AT&T's ownership of Western Electric, the dissolution of Western Electric into three companies, and an end to all restrictive agreements between AT&T, the Bell operating companies, and Western Electric. The focus of the suit was the separation of regulated monopoly services and unregulated equipment supply. That separation had already been imposed in other regulated industries, such as electric utilities, railroads, and airline service. The complaint described the contracts among AT&T, Western Electric, and the Bell operating companies as a price fixing conspiracy and also as an aid to Western Electric's monopolization of the telephone equipment market.[16]

AT&T developed a three-part plan to meet the antitrust threat: (1) denial of the charges and preparation for legal defense, (2) emphasis on the company's defense-related work and enlistment of the aid of the Department of Defense against the Justice Department charges, and (3) negotiation of a settlement restricting the company to regulated activities in exchange for antitrust freedom within the regulated sphere. AT&T's strategy was successful in maintaining the company as a dominant integrated company. It emerged from the confrontation with equal dominance of the telecommunications markets as before, but with a far closer tie to regulation and less freedom to respond to market opportunities than before.

AT&T's legal defense asserted that unified operation of R&D, manufacturing, and final provision of service was necessary for effective telephone service and denied that the arrangements in effect between the various Bell companies constituted a restraint of trade. Bell attorneys emphasized the regulated monopoly character of the company's service and described Western Electric as the "supply arm" of that regulated service. They contended that Western Electric was subject to indirect regulation through the ability of commissions to challenge Western Electric prices in regulatory hearings. The company denied that a market for communications equipment as such existed; thus Western Electric could not monopolize such a market. The company also made a special defense based on the 1932 Consent Decree that required modifications to the patent license agreements among AT&T, RCA, GE, and Westinghouse. AT&T contended that that consent decree implied judicial sanction of the arrangements made up to that time and absolved the company from antitrust liability for the previous patent agreeements.[17]

After filing a formal response to the antitrust charges, AT&T sought to have the suit delayed or dismissed by emphasizing the role of the company in national defense work. AT&T's communications and electronics expertise had made it an important defense contractor during World War II and the early years of the Cold War. The outbreak of the Korean War intensified AT&T's military role and political sensitivity to actions that might adversely impact a defense contractor. AT&T's first opportunity to make a political defense against the suit came four months after it was filed, when AT&T was asked to take over management of the production of atomic weapons at Sandia Laboratories. Sandia had previously been operated by the University of California at Berkeley, but the university wanted to withdraw as Sandia changed from a research to a weapons production orientation.[18] AT&T accepted the assignment and used the occasion to emphasize the importance of the unified Bell system and the dangers of the antitrust suit. The president of AT&T wrote to the chairman of the Atomic Energy Commission:

> The Bell System has always stood ready to do its part in the national defense by undertaking work for which it is particularly fitted; and we agree with you that our organization and methods of opera-

tion give us special qualifications for the work you have described. For these reasons we are willing to undertake the project.

However, as I informed you and your associates who were present at our first meeting on May 30, we are concerned by the fact that the anti-trust suit brought by the Department of Justice last January seeks to terminate the very same Western Electric–Bell Laboratories–Bell System relationship which gives our organization the unique qualifications to which you refer. If Western Electric would enter into a contract to operate the Sandia Laboratories, that company and the Bell Laboratories would indeed work as one, as they now do in Bell System affairs, and the effectiveness of their work would depend, as we have explained to you that it always has, upon their close connection as units of the Bell system. This situation in relation to the anti-trust suit made us question with you whether it would in fact be desirable in the national interest for us to undertake this project, which we agreed if undertaken would probably need to cover a five-year period.[19]

The suit was inactive for two years after the Sandia Laoratories contract, but in 1951 the government began active preparation by sending out document demands and interrogatories to the defendants. AT&T then sought the help of the Defense Department in suspending the suit for the duration of the Korean war to prevent the diversion of AT&T's resources from national defense to the company's defense. The president of Bell Laboratories prepared a memo for the secretary of defense showing that AT&T's existing and planned military work totaled approximately one billion dollars. He contended that if the suit went forward, all of the key executives in Bell Laboratories and Western Electric would spend "substantial portions" of their time on the suit and be limited in their ability to carry out military commitments. The memo prompted the secretary of defense to write to the attorney general as follows:

I understand that recently steps have been taken by your Department looking toward the trial of this suit. These include the service of interrogatories and an order for the discovery of certain documents. The president of the Bell Laboratories has informed me that compliance with these demands and the further activities necessary for a proper defense of this suit are diverting and will increasingly divert key personnel from the work I have outlined above to an extent that will seriously retard the progress we so urgently desire.

Under these circumstances, it is clear to me that the mobilization effort will be impeded by pressing this suit, and that accordingly the interests of the United States require that, without prejudice to the merits for the anti-trust suit, its trial and preparations therefore be postponed while the present situation continues to exist. I therefore request that you grant such postponement.[20]

The attorney general refused to grant a formal postponement but agreed to move slowly on the case in what amonted to a postponement during the election year of 1952. When the new administration came into office in 1953, AT&T sought Defense Department support for its attempt to have the suit dismissed. The company prepared a memorandum for the Department of Defense which was incorporated almost verbatim into a letter from the secretary of defense to the attorney general requesting dismissal of the suit. The letter stated in part:

> Currently, Western has orders for equipment and systems for the armed services totaling over $1 billion. These equipments and systems are the direct result of research and development work performed by Bell Telephone Laboratories in close collaboration with Western, and the laboratories are currently engaged in research projects for this Department at the rate of about $55 million per year . . .
>
> The pending antitrust case seriously threatens the continuation of the important work which the Bell System is now carrying forward in the interests of national defense. This is for the reason that the severance of Western Electric from the system would effectively disintegrate the coordinated organization which is fundamental to the successful carrying forward of these critical defense projects, and it appears could virtually destroy its usefulness for the future. This result would, in the judgement of this Department, be contrary to the vital interests of the Nation . . .
>
> . . .It is now evident that a mere postponement of the prosecution of this case does not adequately protect the vital interests involved. It is therefore respectfully urged that the Department of Justice review this situation with a view of making suggestions as to how this potential hazard to national security can be removed or alleviated.[21]

AT&T's formal request for dismissal of the suit, supported by the Defense Department, did not succeed immediately but did lead to negotiations for a consent decree. The new attorney general took the position that the suit could not be simply dismissed, but that it would be better to settle with a mildly restrictive consent decree than to go to trial. In the summer of 1953, AT&T's general counsel, T. B. Price, found an opportunity for a private meeting with the attorney general at a judicial conference. Price recounted the conversation as follows:

> I then made a number of statements about the injury the case threatened to our efficiency and progress as a communications company and to our contribution to the national defense, told him that under the previous administration we had temporized by asking for postponement only, but that we were hopeful that he would see his way clear to have the case dropped . . .
>
> He reflected a moment and said in substance that a way ought to be found to get rid of the case. He asked me whether, if we reviewed our practices, we would be able to find things we are doing which were once considered entirely legal, but might now be in violation of the antitrust laws or questionable in that respect. He went on to say that he would be surprised if in a business such as ours things of this sort did not exist. The interpretation of the antitrust laws had changed and the courts had very different views today about many business practices from the earlier opinions. Consequently, he thought that we could readily find practices that we might agree to have enjoined with no real injury to our business . . .
>
> As I got up to go he walked down the steps with me and repeated his statement that it was important to get this case disposed of. He said the President would understand this also and that if a settlement was worked out he could get the President's approval in 5 minutes.[22]

After the attorney general's suggestion that a settlement be worked out, negotiations were begun that led to a Consent Decree in January 1956. The Consent Decree effectively accepted the AT&T view of the suit that "the basic issues it raises are matters for legislative and regulatory policy; they have been so recognized by Congress and are being appropriately dealt with; and the suit seeks to have the court take action in conflict with legislative and regulatory treatment in a highly technical field."[23] The decree allowed the existing arrangements among the various companies of the Bell System to continue. Its two substantive provisions restricted the scope of Bell System activities and required liberal licensing of Bell System patents.

AT&T was "enjoined and restrained from engaging, either directly, or indirectly through its subsidiaries other than Western and Western's subsidiaries, in any business other than the furnishing of common carrier communications services" with minor specified exemptions. Similarly

Western Electric was "enjoined and restrained from engaging, either directly or indirectly, in any business not of a character or type engaged in by Western or its subsidiaries for Companies of the Bell System" with certain exceptions including the right to make any equipment for the United States government. The restrictions confined AT&T to regulated services and Western Electric to making equipment for those regulated services. The restrictions had little effect on the then current operations of the companies because very little of its business was outside the common carrier area. However, the decree prevented expansion to serve new unregulated markets and produced a far closer tie between AT&T and the regulatory authorities than had previously existed. Because AT&T was restricted to regulated activities, it became dependent on regulatory decisions and lost the freedom to enter other markets when it saw that as advantageous. Without being explicit, the decree also gave some credibility to AT&T's contention that its regulated status exempted it from the antitrust laws by releasing the company from antitrust liability and confining it to regulated activities. AT&T's conception of Western Electric as a captive supply arm of a regulated utility rather than a general manufacturing company that monopolized the telephone equipment market was endorsed by allowing the continuation of AT&T's ownership of Western Electric, but restricting Western Electric to producing the type of equipment used by Bell operating companies.

The second substantive provision of the decree was a requirement that all patents controlled by the Bell System be licensed to others on request. Licenses under the 8600 patents that were involved in the cross-licensing agreements among AT&T, General Electric, RCA, and Westinghouse were required to be issued royalty free to any applicant except the parties to the cross-licensing agreement. Licenses under all other existing or future patents were required to be issued to any applicant at a "reasonable royalty" with provision for the court to set the royalty if the parties could not agree. AT&T was also required to provide technical information along with patent licenses on payment of reasonable fees.[24]

The patent license provisions required some liberalization of the Bell licensing policy. After World War II, Bell had relaxed its stringent requirements for licensing patents to others and had licensed many patents, including the vital transistor patents, before the Consent Decree. However, the decree made many of the most important patents royalty free and restricted AT&T's freedom to determine the conditions under which other patents would be licensed. The licensing provisions were in accord with the conception of the Bell System as a regulated monopoly rather than a market-controlled company. Within the regulated arena, the regulators controlled entry "in the public interest" and thus there was no need for protection against entry from the patents. Because Bell could not enter unregulated businesses according to the Consent Decree, compulsory licensing of the patents was a method of making the technology available where the Bell System could not exploit it. The government retained the right to reopen the case if regulation was discontinued by a substantial number of states. AT&T analyzed the "General Theory" of the consent decree as follows: "Regulatory processes adequately protect the public interest in the field of common carrier communications services, including manufacture by Western of equipment designed for such use."[25]

The significance of the licensing provisions for actual Bell practice at the time may be evaluated by comparing the transistor licenses before and after the decree. At the time of the final judgment, Western Electric had fifty-nine transistor licensees. Each licensee paid an advance fee of $25,000 which was credited against royalties. Royalties were fixed at 5 percent of the net selling price before July 1, 1953, and at 2 percent after that date. The license agreements were for a minimum period of four years after which the agreement could be terminated on one year's notice by either party. The license agreement as such did not give the licensee the right to technical information, but as a matter of practice the licensees were given the technical information at symposia of Bell Laboratories to which they were invited without additional payment. After the Consent Decree, transistor licenses were granted royalty free under the patents existing at the time of the judgment and at a royalty rate of 2.5 percent of the selling price for transistors using patents issued after the date of the decree. The royalty could be adjusted downward according to the value of licenses granted back to Western Electric by the licensee. No technical information was included in the licenses. For a fee of $25,000, Western Electric would furnish to royalty-free licensees the technical information which was previously furnished free to royalty-paying licensees. Other technical information would be charged for at the cost of gathering it plus a percentage fee based on the value of transistors manufactured by the licensee.[26]

The total effect was not greatly different than the situation before the Consent Decree except that the licensee had a legal right to the license and technical information and was not dependent on potential changes in Bell policies. The legal right to produce transistors was significant for investment decisions. The pre-decree policy allowing termination after four years could hamper significant investment in R&D or manufacturing facilities because of the possibility that they could not be used until the basic transistor patent expired if the Bell System changed its licensing policy. Under the Consent Decree, companies that were sophisticated enough to develop their own manufacturing processes and semiconductor improvements were freed from dependence on the basic patent and could proceed royalty-free. Companies that wanted to take advantage of continuing Bell System advances in either basic research or manufacturing could do so by the payment of royalties. The transistor license agreements caused a great upsurge in transistor manufacturers and a rapid drop in transistor prices.

The Consent Decree marked an important change in the Bell System philosophy. For the first eighty years of the system, it put great emphasis on patents. Patents were used to establish the monopoly, prevent entry, and establish industry boundaries. The Consent Decree gave the system the right to continue dominance of the industry through regulatory methods, but reduced its patent control. From that time on, the system would look primarily to regulatory and legislative protection for its barriers to entry. While the Consent Decree did not make major changes in existing practices of the Bell System or require any significant divestiture, it did accelerate the transformation of AT&T from a private corporation to a semipublic regulated corporation.

Evaluation of the Regulated Monopoly Era

Regulation helped maintain AT&T's market dominance against the threats of technological change and antitrust action during the 1934–1956 period. In the case of microwave, the inherently slow procedures of regulation prevented new entrants from exploiting the profitable opportunity caused by the simultaneous advent of new technology and a new source of demand. The key initial fact in dispute was the demand and supply for microwave frequencies. If an adequate number of microwave frequencies existed relative to the common carrier demand for them, an open entry policy would promote the rapid development and use of the new technology. If microwave frequencies were very limited, then a restrictive licensing policy would be desirable to avoid tying up scarce frequency allocations by private companies that might not fully utilize them. The initial frequency congestion in AM radio broadcasting prior to regulation gave credence to AT&T's assertion that microwave frequencies were scarce and should be reserved to the common carriers. While other companies disagreed with the assertion, the commission lacked any method of gathering independent facts that would clearly indicate the range of frequencies practical for microwave use and the demand for those frequencies by common carriers. Lacking adequate data to make a firm policy decision, the commission initially delayed resolution of the factual issue of the availability of frequencies and granted only experimental licenses.

The decision to grant experimental licenses required potential entrants to take the risks of developing a new technology without the assurance that they would be granted the legal right to continue in business if the technology proved to be commercially feasible. The new companies could have been expected to move rapidly to take advantage of the excess demand for communications services after World War II if there had been no FCC restrictions. With only experimental licenses and the knowledge that AT&T had the ability and the incentive to at least delay the granting of regular licenses, the new companies had no incentive to invest significant amounts in building microwave networks.

The FCC decision to grant regular microwave licenses only to common carriers, but only to require interconnection with non–common carriers eliminated the threat of microwave competition to AT&T. The commission's refusal to grant regular licenses to non–common carriers can be interpreted as a conservative response to inadequate information about the demand and supply of frequencies. So long as there was a possibility that inadequate frequencies were available to meet the needs of regulated common carriers, the commission could not be expected to grant some of the frequencies to companies outside of its control. As better information on the availability of frequencies came to the commission's attention during the 1950s, it relaxed the rules and licensed companies other than common carriers.

The commission's reluctance to require interconnection among common carriers is more difficult to understand. Interconnection of telephone companies had been a well-established principle since the Kingsbury Commitment of 1913 when Bell had agreed to interconnection under antitrust pressure. Prior to the formation of the Federal Communications Commission, interconnection had taken place either voluntarily or as a response to antitrust action, rather than being imposed by the Interstate Commerce Commission. The granting of explicit authority to require interconnection to the FCC appeared to strengthen AT&T's obligation to connect with other companies, but instead was used to negate the previous requirements. The FCC focused on the language in the law granting the commission authority to require interconnection after a hearing and a finding that interconnection was in the public interest. It interpreted the language to require a specific hearing on the facts involved in each route, rather than a general policy procedure that declared interconnection promoted the public purpose. The courts refused jurisdiction to consider the previous antitrust interconnection requirements so long as the matter was before the FCC. Thus the delayed hearing resulted in no legal requirement for interconnection during the crucial period. Because the specific interconnection issue was between Western Union and AT&T, both firms regulated by the commission, the theory that a regulatory commission protects its firms from outside competition does not adequately explain the commission's action. It is unclear whether the commission was consciously attempting to protect AT&T from Western Union competition or was simply being very conservative in its interpretation of the law and attempting to protect itself from being overruled by an appeals court.

The FCC actions not only protected AT&T's market position, but also slowed the spread of microwave technology and extended the period of excess demand for television network capacity. Without the FCC intervention, microwave networks probably would have spread rapidly during the late 1940s. Networks could have been built by private users (including broadcasters and businesses with large volumes of telephone traffic) or common carriers. The various networks could have then been interconnected using antitrust precedent without specific regulatory requirements. Even with the restrictive FCC policies, many broadcasters found it profitable to build temporary microwave networks because of their inability to obtain service from AT&T. The broadcasters could have been expected to build more extensive networks if they had not been required to abandon their private networks when AT&T facilities became available. It is also likely that AT&T would have moved more rapidly to satisfy the excess demand for communications facilities if potential entrants were allowed to satisfy the demand than it did in the actual situation in which it was guaranteed the business when it completed its facilities. Thus the social effect of the FCC actions was to delay the introduction of microwave technology and extend the period of excess demand beyond what it would have been under free-market conditions.

The antitrust suit was an attempt to impose changes on AT&T's structure that the FCC had considered but not formally requested. The Communications Act of 1934 had foreseen possible legislative changes in the Western Electric–AT&T tie in its order for the commission to investigate the industry and request further legislation as necessary. Separation of ownership between equipment manufacturing and regulated utility service was imposed legislatively on other industries during the 1930s and considerable sentiment existed for similar treatment for communications. After failing to secure FCC sanction for their recommendations, some of the investigation staff joined the Justice Department and led preparations for the antitrust suit.

AT&T was able to blunt the potentially procompetitive nature of the antitrust suit by appealing to its status as a regulated monopoly of national importance. Rather than attempting to portray itself as a firm surrounded by competitors, as often happens in antitrust cases, AT&T emphasized its technological skills and its control by regulation instead of competition. AT&T's importance as a military contractor during the Korean War and the Department of Defense's support of the company's assertion that its integrated organization was necessary to the maintenance of its technological expertise made it difficult for the Justice Department to be certain that the suit was promoting the public interest. The Consent Decree's restriction of AT&T to regulated activities essentially accepted the company's position that regulation was a substitute for antitrust and that a regulated firm should not be subject to the antitrust laws. AT&T gave up the ability to protect its market power through patents and entry into other markets but gained the freedom to continue operating as a dominant integrated firm in the regulated markets. It thus became dependent on regulatory decisions for protection from competitors and for the scope of its activities. Given AT&T's historical experience with antitrust and regulation, placing itself under the protection of the regulatory commissions in exchange for dropping the antitrust suit appeared to be a clear gain for the firm. Later decisions by the regulators made the benefits to AT&T from regulation less certain than they appeared in 1956.

8 New Competition I: The Long-Distance Market

The Long-Distance Market in 1956

The barriers to entry into long-distance service in 1956 were formidable. Although no explicit monopoly had been granted by the Federal Communications Commission, the commission's acts had provided substantial legal protection from entry. The most feasible technology for potential entrants was microwave, which required a license from the FCC because of its radio nature. The FCC's previous treatment of broadcasters' license requests suggested that licensing would be a significant barrier for a potential entrant. If the potential entrant also wanted to operate as a common carrier rather than only provide services for its own use, further FCC approval was necessary.

From AT&T's point of view, the regulatory barrier to entry was superior to market barriers. With favorable agency decisions, regulation could be an absolute bar to entry regardless of the incentives to enter. Even with adverse decisions, the procedural requirements of regulation provided ample opportunity to delay any potential entrant for several years. The delay would not only protect against entry during the delay time, but would increase the costs of the entrant by requiring expenditures to be continued for several years prior to the receipt of revenue. In addition, the delay would provide time for the dominant firm to prepare a market response if entry did occur.

If legal barriers to entry failed, AT&T's next line of defense was interconnection restrictions. Although interconnection requirements had been established via antitrust pressure prior to the formation of the FCC in 1913, the FCC's rejection of Western Union's request for interconnection of television facilities suggested that interconnection was again an allowable competitive weapon. Interconnection would be particularly useful against an entrant attempting to provide only long-distance service because AT&T controlled the operating companies for most of the phones and could directly prohibit local-service connections. Because the long-distance service would be useless without some method of connecting to actual subscribers, the control of local connections could provide an absolute bar to competitive entry in long-distance service if no regulatory restrictions were placed on the use of interconnection.

The third barrier to entry was economies of scale. Economies of scale in microwave transmission were very strong up to about 240 voice grade circuits, moderate between 240 and 1,000, and insignificant at levels above 1,000 circuits. The basic components of a microwave system were property (tower and storage building every thirty miles along the route), radio equipment (the actual equipment for receiving, amplifying, and transmitting the microwaves), and multiplex equipment (the equipment to impose multiple telephone conversations on a single microwave radio transmission). The property and radio costs were approximately constant up to about 1,000 circuits (the capacity of one microwave transmission), while the multiplex costs were almost linearly related to the number of circuits the system was designed to handle. At low levels the costs were dominated by property and radio, while at high levels the costs were dominated by multiplex equipment. One study showed the cost of multiplex equipment increasing from 22 percent of the total cost of a microwave system of 24 circuit capacity to 79 percent of the total cost of a system of 996 circuit capacity.[1]

At low levels, the economies of scale were substantial. Figure 4 shows the cost per voice circuit per mile per month for a 100-mile-long microwave system at various levels of circuit capacity according to figures from AT&T and Motorola. Although the two studies differ in the absolute magnitude of the figures, they both show significant economies of scale in the range of 12 to 240 circuits. The significance of these economies of scale varied with the density of the market. In low-density markets, the entire communications demand was less than 240 circuits and the economies of scale were a substantial barrier to entry. In high-density markets (between the largest cities), the total demand was several

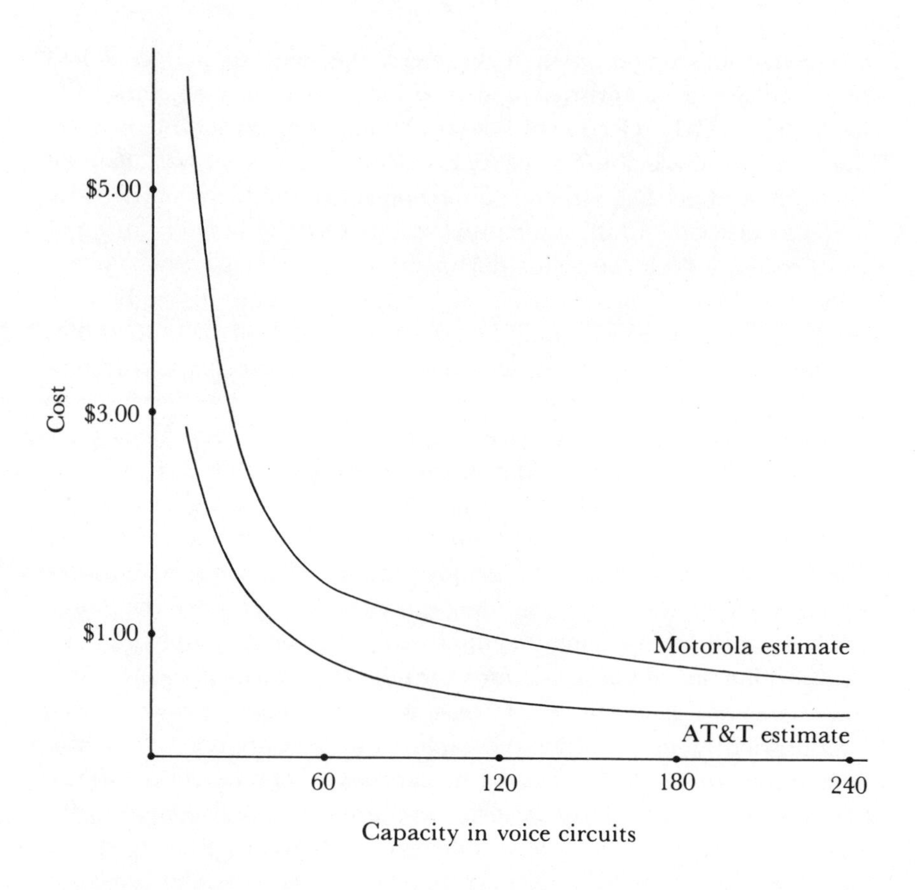

FIGURE 4 Cost per circuit per mile per month for a 100-mile microwave system (data from Telpak Tentative Decision, 38 *F.C.C.* 370 at 385, 386)

thousand circuits and economies of scale were a much less serious barrier to entry. Consequently the economies of scale were not so great as to create a traditional "natural monopoly" but did restrict potential entrants to the high-density routes.

Although the total barriers to entry were very high, AT&T's pricing policies created a strong incentive to enter. The overall rate level for long-distance services was above the competitive level. No legal restrictions had been put on AT&T's interstate rate of return either by the ICC or the FCC. Because of rapid technological progress, the company was able to avoid requesting rate increases and voluntarily make some reductions. Consequently there was no pressure on the FCC to investigate how the rates compared to the competitive level. The rate was above the level necessary to attract entry, given the assumption that only economies of scale were the relevant barriers to entry. It was below the pure monopoly level computed on the assumption that the FCC would guarantee AT&T's monopoly.

Additional entry incentives came from the rate structure. Rates were based on a uniform nationwide charge per circuit for a given mileage. No volume discounts were available. Thus a large-scale user on a high-density route paid the same rate per circuit mile as a small business in a lightly populated area. Rates were based on average costs of installed equipment rather than current costs of the most efficient equipment. AT&T's "cost" for communication between New York and Philadelphia was not the least cost current technology but the average of microwave, coaxial cable, and older open-wire systems still in use. The rapid technological progress in long-distance communications combined with AT&T's slow depreciation of installed equipment caused AT&T's book cost of circuit miles to be above the level of minimum cost technology. In a competitive market, the old equipment would have suffered a capital loss as cheaper equipment became available, but in the monopoly market the old equipment could still be carried at original book values as if technological progress had not taken place.

A further factor which increased the incentive of new firms to enter the market was the practice of allocating some of the costs of local telephone service to long distance through separations formulas. A long-distance call placed from one local telephone company to another would go through the local lines of the first company, the long-distance lines of AT&T, and then to the local lines of the receiving company. The originating company would bill the customer for the call and the revenue would be divided between the local company and AT&T. Separations only applied to dial-up calls, not to private-line service which provided a long-distance circuit for the exclusive use of one customer. The separations formulas were worked out in conferences of the telephone companies in cooperation with state regulatory agencies. The general principle used was that costs should be allocated according to relative use. The average commitment of the local telephone lines to long-distance service was considered a claim on long-distance funds for that percentage of total local costs. Over time, the formulas assigned a greater and greater portion of the long-distance revenue to connecting local service. From AT&T's point of view, much of the separations money was simply internal accounting transactions between the Long Lines Department and the operating companies, with a small percentage of the separations money leaving AT&T control to pay independents for their local wires. Even payments to the independents benefitted AT&T to some extent because the dependence of the independent telephone companies on separations payments from AT&T gave them a strong community of interest with AT&T and caused them to support AT&T's continued monopoly of long-distance service.

The state regulators generally supported increased assignment of the local exchange plant to long distance for separations purposes because that reduced revenue requirements for local service. The FCC had exactly the opposite incentives; but, because long-distance rates were falling anyway and the FCC was not being asked for rate increases, it took no part in the separations negotiations until after 1965. Consequently, instead of the true cost-based situation of rapidly falling long-distance rates and rising local rates, changes in the separations formulas caused the long-distance revenue to increasingly subsidize the local revenue and led to slowly falling long-distance rates and slowly rising local rates with greater and greater divergence between true costs and rates.

The entry incentives created many potential entrants into the industry if the regulatory barriers to entry could be overcome. As entry was attempted, AT&T sought protection from each of the three barriers to entry: regulatory protection, interconnection restrictions, and economies of scale in low-density markets. The combination of responses was enough to keep the total sales of the entrants to a miniscule level, but the attempted entry caused significant changes in the character of regulation and the pricing policies of AT&T.

Private Communications Systems and Response

At the end of 1956, the FCC undertook a review of its microwave frequency allocations. The review included reconsideration of allowing private users to construct microwave systems for their own use when common carrier facilities were available. The existing policy at that time was to grant microwave licenses only to common carriers (AT&T and Western Union), television broadcasters who could not obtain common carrier service, and right-of-way companies (pipelines and railroads).

That policy was based on the presumed shortage of frequencies in the microwave band and the assumption that the limited frequencies would be used most efficiently by common carriers.

Considerable technological progress in microwave technology and the increasing use of microwave for television and voice transmission made the review necessary. Technological progress after the initial allocations had provided the ability to use a much greater range of frequencies for microwave communication and allowed microwave system designs that made more intensive use of the available frequencies. At the time of the hearing, microwave had come into extensive use. It accounted for 22 percent of the 10.5 million circuit miles of telephone capacity and 78 percent of the 60,000 circuit miles of intercity television capacity in the Bell System.[2] Western Union was continuing to use its limited microwave system between New York, Washington, and Pittsburgh for telegraph service. Private microwave systems to serve right-of-way companies included 14,000 route miles for utilities and 17,000 miles along petroleum pipelines.

Extensive hearings were held in which over two hundred persons or organizations participated. The opinion split clearly into two groups. The common carriers (AT&T, Western Union, and the independent telephone companies) contended that frequencies were scarce and should be reserved to common carriers, and that in any case the commission had a duty to protect the carriers' economic interest by prohibiting competition. The common carriers took the position that only public safety and right-of-way companies should be allowed private microwave, that sharing of a system should not be allowed, and that private systems should only be licensed when common carrier facilities were not available and should be abandoned when common carrier service became available. According to the FCC report:

> The Bell System witnesses contended that to permit the licensing of private systems where common carrier facilities are available would cause irreparable harm to the telephone company's ability to provide a basic nationwide communication service, which is vital in times of peace but indispensable in times of national emergency. Also, they claimed that widespread licensing of private systems would not only increase the cost of communications to the Nation's economy as a whole but would cast an added burden upon the individual and the small businessman who would continue to rely on common carriers.[3]

In contrast, the potential private users argued that there should be no limitations on eligibility to establish private systems, that sharing of a system by more than one user should be allowed, and that the availability of common carrier facilities should not be a factor in licensing private systems. The microwave equipment manufacturers and potential users of private systems contended that, while frequencies were not unlimited, there were adequate frequencies available for both common carriers and private users and that the commission had no duty to protect the economic interests of the carriers. The Electronic Industries Association (representing microwave manufacturers) presented an extensive engineering study which indicated that 742 microwave stations could be accommodated in a city terminal area without interference using existing technology and frequency bands, and that greater concentration could be obtained by special engineering procedures. At that time, Los Angeles had the greatest number of microwave stations with thirty-eight, followed by thirteen in San Francisco, nine in Phoenix, and fewer in other cities, leaving room for extensive expansion before interference would occur.[4] The Electronic Industries Association study provided critical factual information on frequency availability that was unavailable to the FCC at the time of the restrictive 1948 policy. The study represented six thousand man-hours of effort and indicated a significant financial investment by the association in reversing the restrictive licensing policy. Although AT&T challenged the validity of the study, it was given considerable weight by the commission.

In July 1959 the FCC announced its conclusion that:

> there are now available adequate frequencies above 890 mc. to take care of the present and reasonably forseeable future needs of both the common carriers and private users for point-to-point communications systems, provided that orderly and systematic procedures and technical criteria are applied in the issuance of such authorizations, and that implementation is consistently achieved with respect to all available and future improvements in the art.[5]

This finding was a reversal of the factual assumption underlying the previous microwave restrictions and led the FCC to make microwave available to any private user. The commission explicitly ruled that the availability of common carrier facilities would not be a factor in granting private microwave applications. However, the commission declined to authorize the sharing of private systems and thus restricted the new freedom to companies with very dense communications demand between two points.

The FCC avoided an explicit ruling on several important issues considered during the hearing. On the question of interconnection between private systems and common carriers, the FCC ruled that the requirement to interconnect should be determined on a case-by-case basis rather than in a general proceeding. On the question of the FCC's responsibility to protect the economic interests of existing carriers, the commission ruled that it had no explicit duty to do so, but that it would defer a ruling on its implied responsibility to do so because of the absence of a showing that significant economic harm would come to the existing carriers from the proposed private systems. Thus the commission left open the possibility that future competition could be curtailed in order to avoid economic harm to existing carriers. Given the commission's factual assumptions of adequate frequencies and no harm to existing carriers, the decision protected the interests of all parties. It was not designed to create competition or to harm AT&T but merely to allow private users the benefit of otherwise idle frequency bands.

The July 1959 decision allowing private microwave did not go into immediate effect because of petitions from the common carriers requesting reconsideration of the decision. The petitions reargued many of the issues previously considered in the hearing and also brought up the new issue of frequency allocations for space communication. The Russian *Sputnik* in 1957 and U.S. *Explorer* in 1958 had shown the feasibility of artificial satellites and implied a potential new use for microwave frequencies for communication via satellite. AT&T argued that satellite frequency needs would be so great as to invalidate the commission's finding that adequate frequency space was available for both common carriers and private microwave systems. While the carriers could not specify the frequencies needed for space communications, they argued that any allocations to private users should wait until space needs became known. Consequently the commission ordered a stay in implementation of the decision and reopened the record for additional information regarding satellite communications. A new opinion was issued in October 1960 essentially reaffirming the original decision and rejecting AT&T's assumption regarding satellite frequency requirements.[7]

The FCC decision to license private systems was not a declaration of open entry into long-distance communications. It did not authorize new companies to enter the business in competition with AT&T but only allowed individual users to build systems for their own use. It was similar to the Interstate Commerce Commission's provision for corporations to use their own trucks without common carrier regulations or authorization. However, the exception was more narrow than the private trucking exception to regulation because of the economies of scale of microwave systems and the lack of interconnection requirements. No provision was made to connect the private systems into common carrier systems. Consequently a New York company with extensive business in Boston spread among many different Boston companies would not find a private system from New York to Boston feasible, because it would need to connect its system to the Boston local system in order to reach its customers. However, a New York firm with a manufacturing plant in Boston could use a private system for communications between its New York office and its Boston plant. Thus, despite the FCC relaxation of the regulatory protection, the interconnection and sharing restrictions limited the potential entrants to firms with bulk communications needs between two distinct points.

At the time of the decision, AT&T's charge for a 100-mile private line was $315 per month per line. The private line then gave the customer unlimited use between those two points as if they were in the local telephone area. Multiple lines were charged for at multiples of the single-line rate. Motorola's estimate of the monthly cost of owning and operating a 100-mile private microwave system equipped for twenty-four voice channels was $7,707, equal to the cost of 24.5 private lines from AT&T. For 240 lines, a 100-mile private system would cost $17,141 per month while the equivalent AT&T lines would cost $75,600 per month.[8] The private system would need to be substantially cheaper than the Bell rates in order to be competitive because of the lower flexibility of the private system, the inability to reconfigure it quickly if needs changed, and the potential administrative problems of running one's own communications system. Thus the existing rates made it unlikely that users with fewer than about fifty lines would build a private system but provided strong incentives for high-volume users to set up a private system.

The FCC decision made one very small part of the long-distance market vulnerable to competition at the existing prices. AT&T then faced the strategic decision of whether to deter entry through pricing action or to allow entry. Because potential entry was limited to a small part of the long-distance market, an across the board price cut would have been a very expensive way to deter the new competition. However, if a price discrimination scheme could be developed to give lower prices to those likely to build private microwave systems while retaining the original prices for those unable to find alternatives, entry could be deterred at low cost. If the entry deterring prices were above cost, deterrence via price discrimination would clearly increase profits because the entry deterring prices would save some profits that would otherwise be lost. If the necessary entry deterring price was below AT&T's cost, it could still be a profitable action if it prevented later entry in markets not threatened at that time. The existence of private microwave networks could provide potential entrants into other segments of long-distance communications if the FCC should so allow. In addition, the actual existence of private microwave networks would make FCC decisions allowing greater entry more likely because of the ability of those firms to present a case in favor of entry. So long as price discrimination could be used to distinguish between customers likely to build private microwave systems and those unlikely to do so, a selective price reduction to make entry uneconomical would be beneficial to AT&T.

Four months after the final decision allowing private microwave systems, AT&T filed a price response that largely eliminated the incentive for any company to build a private system where AT&T facilities were available. The new tariff, called Telpak, offered large discounts for groups of private lines. The Telpak tariff offered a discount of 51 percent for twelve lines (Telpak A), 64 percent for twenty-four lines (Telpak B), 77 percent for sixty lines (Telpak C), and 85 percent for 240 lines (Telpak D). Thus, for a set of 240 lines over a distance of 100 miles, a customer would have paid $75,600 per month under the old rates and $11,700 per month under the new rates.[9] The new charge was below Motorola's estimate of $17,141/month for a private system. The rates for a single private line were not changed.

Because the new microwave systems were to be built by individual users rather than companies seeking microwave business, there was no opposition to the Telpak tariffs from the companies planning microwave systems. The users' goal was inexpensive bulk communications service and if AT&T would provide it more cheaply than they could provide it for themselves, they were happy to continue as AT&T customers. Consequently most of the potential users of the newly authorized private microwave systems willingly abandoned planning for their own systems, signed up for Telpak, and supported the importance of Telpak in subsequent hearings. However, Western Union and the potential suppliers of microwave equipment saw the Telpak tariff as a discriminatory and predatory pricing scheme which would take away Western Union's private-line business through unremunerative AT&T rates subsidized by its monopoly services, and would stunt the non-Western Electric market for microwave equipment.

At the request of Motorola (a microwave equipment manufacturer) and Western Union, the Telpak tariff was suspended for ninety days (the maximum period allowed by law) and an investigation into its lawfulness was begun. Extensive hearings were held before a decision was issued in December 1964, nearly four years after the original filing of the tariff. The commission found that the Telpak lines were indistinguishable from ordinary private lines purchased one at a time and that there was no cost savings to AT&T from selling them in bulk. The same facilities were used as for single private lines and the potential savings in accounting and administrative expense from selling in bulk only compensated for the expense of the complex accounting procedures for Telpak lines. The tariff was judged to be a case of price discrimination without cost or service differences to justify it. This was contrary to the provisions of the Communications Act, unless it was found in addition to be both compensatory and justified by competitive necessity. Telpak A and B (12 and 24 circuits) were found not to be justified by competitive necessity because it was infeasible to build a private microwave system for such a small number of circuits. Those tariffs were canceled. Telpak C and D (60 and 240 circuits) were found to be justified by competitive necessity because the old rates could not compete with private microwave. However, the commission did not have enough data to decide whether or not the rates covered costs. The commission consequently resumed the hearings to allow AT&T to present additional cost information on Telpak C and D. The tariffs continued in effect while the question was being considered.[10]

The Telpak case appears destined to provide lifetime careers for its antagonists. The post-1964 proceedings examined the relationship of Telpak C and D prices and costs. That question could not be answered without specifying accounting conventions in considerably more detail than the commission had previously. Telpak used the same facilities and personnel as other services and many possible methods of apportioning the costs existed. Some methods showed Telpak prices above costs and some did not. The Telpak hearings were combined with other issues related to cost allocation. Many years of hearings were held. Finally in

1976 the commission ruled that Telpak C and D were illegal and ordered the tariffs withdrawn by June 1977. During the fifteen years of Telpak consideration, many users (including the United States government) had developed private communications systems based on Telpak rates. Because the rates were lower than any alternatives, they strongly supported the continuation of Telpak rates. After the adverse 1976 decision, Telpak users successfully petitioned for reconsideration on the grounds that inadequate notice of the issues in previous proceedings had been provided. The order to withdraw Telpak was rescinded and a new hearing on the lawfulness of Telpak was ordered.[11]

Meanwhile, in a separate proceeding the commission was considering the lawfulness of the restrictions imposed by AT&T on the resale or shared use of its services. It decided that such restrictions were illegal.[12] After the commission's 1977 order to eliminate resale restrictions, AT&T sought a court stay with regard to Telpak, because resale would make the reduced prices available to small users as well as large. AT&T obtained a short stay and then voluntarily canceled the Telpak tariff at the expiration of the stay (June 1977). The Telpak users then sought an injunction against AT&T's cancellation of the Telpak tariff because it constituted elimination of an existing service or a massive and unjustified price increase. The FCC refused to challenge the voluntary elimination of Telpak, but the users obtained an appeals court order maintaining Telpak in effect for existing users while litigation continued.[13] In oral argument before the appeals court in 1979, the FCC contended that there was no proper legal issue because "the carriers are not required to justify non-rates. This is a non-event."[14] The Department of Justice, representing government using agencies, challenged the elimination of Telpak as imposing an $80 million/year rate increase on the government alone. At the end of 1979, no decision had been reached. As the Telpak case finishes its second decade, the tariff continues in effect in spite of the FCC's finding that it was an illegal, discriminatory, below-cost offering and AT&T's desire to voluntarily eliminate it. Once the tariff was instituted by AT&T, users established property rights in it which prevented its withdrawal even with the agreement of the FCC and AT&T.

The result of the initial Telpak hearings was ideal from AT&T's point of view. The company presumably would have preferred to only offer the volume discounts to large users who were likely to build private microwave systems, but offered the smaller discounts for 12 and 24 lines in order to make a defensible volume discount plan. Thus the order to drop the discounts for 12 and 24 lines combined with permission to continue the discounts for 60 and 240 lines (the classes of customers likely to build private systems) allowed the company to deter entry while giving up a minimum of revenue from customers who had no alternative to AT&T service. The extensive proceedings before the FCC prevented court challenges of the FCC's decision or separate antitrust challenges to the tariff (on either price discrimination or monopolization grounds) because the matter was under consideration by the FCC. Thus a pricing revision with a small revenue impact on AT&T was able to prevent entry of private microwave systems and maintain monopoly control against the first FCC relaxation of regulatory barriers to entry.

An AT&T study of the rate of return on various long-distance services for the year ending August 1964 showed that Telpak was the least profitable with a rate of return of 0.3 percent while the main long-distance services (WATS and switched long-distance) earned over 10 percent on capital utilized.[15] Although the Telpak return was far below the cost of capital, the service was such a small percentage of total revenues that the losses on Telpak were a modest investment in preventing entry.

The Entry of Specialized Common Carriers

The Telpak rate structure clearly indicated that the AT&T single line rates were far above microwave costs. If a company could purchase a set of 240 lines at the Telpak bulk discount rate and resell them to individual users who had been paying the single line rate, substantial profits could be made. The AT&T tariff prohibited the resale of Telpak lines or the tariff would have not been viable. All price discrimination schemes require some method of preventing arbitrage between the favored and disfavored customers. The ability to prohibit resale via tariffs was an effective method of accomplishing this in the Telpak case. However, because the Telpak tariff was set up to be competitive with private microwave costs, it should have also been possible for a company to build its own facilities and offer private lines below the AT&T rate if regulatory approval could be gained.

A new entrant would need authorization from the FCC to operate as a common carrier in competition with AT&T. It would also need to secure interconnection arrangements with the local telephone companies in order to provide local distribution of the long-distance services. Because the FCC had not made an explicit monopoly grant to AT&T and because AT&T and independent telephone companies were interconnected, securing the right to operate and interconnect was a possible but very risky undertaking. AT&T could be expected to oppose such a proposal vigorously in the regulatory process. Consequently a new entrant would have to be prepared to incur substantial expenses over several years for the uncertain prospect of the legal right to operate, and then face uncertain AT&T pricing responses if entry rights were obtained.

In response to the entry incentives, Microwave Communications Inc. (MCI) filed a request for authorization as a common carrier in 1963. MCI proposed to build a microwave system between St. Louis and Chicago to offer voice, data transmission, facsimile, and other services over private lines in competition with AT&T. The application was opposed by AT&T, General Telephone, Illinois Bell, Southwestern Bell, and Western Union, who jointly complained that the service would meet no demonstrated need, that it would result in wasteful duplication of facilities, that MCI did not possess the legal, technical, and financial qualifications to install and operate the system, and that the system would cause radio interference to existing and future facilities of the established common carriers. The carriers also asserted that the authorization should be denied as serving no useful purpose, because MCI would be unable to supply local service and the carriers would refuse to provide local service for MCI originated communications.

In February 1966 the FCC ruled that MCI was technically, legally, and financially qualified to undertake the services but that the objections by the carriers raised questions of fact that required a hearing before the commission could make a finding that the applications would serve the public interest, convenience, and necessity.[16] The evidentiary hearings were held in early 1967. In October of that year the hearing examiner recommended that MCI's applications be granted. At the request of the established carriers, the decision was referred to the entire commission for further argument and review. In August 1969 the commission affirmed the examiner's decision on a 4–3 vote.

The three dissenting commissioners viewed the case as a challenge to the established system of regulated monopoly communications services and nationwide rate averaging. They saw the case as prejudging major policy issues of interconnection and rate structure under the guise of a limited hearing for a small-scale construction application. Thus their opposition was based on a correct forecast that the MCI case provided an important precedent and was much more significant than the actual construction applications would suggest.

The majority emphasized the actual facts of the MCI applications rather than the significance of the decision for future policy. The proposal was to build a single-channel microwave over the single route of St. Louis to Chicago. The expected initial service was 75 voice circuits and expected maximum service was 300 voice circuits. The estimated cost of the system was $564,000, most of which would be financed by the equipment supplier.[17] Thus the proposed capacity of the new competitor was no more than would have been expected from a single private user operating under the 1959 private-user authorization.

The majority was unwilling to deny what in effect was a shared private system that provided small users with the benefits of private system costs. An additional factor influencing the majority decision was MCI's proposed flexibility in providing services tailored to individual customer requirements compared with AT&T's tariff rigidity. The commission had been receiving complaints from computer service companies and others about inadequate communications services to meet their specialized needs. Although the MCI service was essentially a price cut on ordinary private-line service, it was packaged as a more flexible service to meet specialized needs. A final factor influencing the MCI approval was growing dissatisfaction with the regulatory process as a whole and an attempt to experiment with alternatives. Commissioner Nicholas Johnson explained his support for the MCI application as follows:

> No one has ever suggested that government regulation is a panacea for men's ills. It is a last resort; a patchwork remedy for the failings and special cases of the marketplace . . . I am not satisfied with the job the FCC has been doing. And I am still looking, at this juncture, for ways to add a little salt and pepper of competition to the rather tasteless stew of regulatory protection that this Commission and Bell have cooked up.[18]

The commission's focus on the specific applications rather than the broader policy issues raised by new competition allowed it to experiment without determining an explicitly procompetition policy. The FCC dodged the issue of whether or not potential economic harm to established carriers was grounds for denying new competition by noting that the proposed MCI facilities were too small to have any effect on AT&T's financial results. Rather than fully considering the interconnection issue at the MCI hearing, the commission deferred the interconnection arguments until actual requests were in place. However, it did announce that the carriers would be expected to provide local loop service unless they could prove that such service was infeasible. Thus the decision gave MCI the legal right to operate but left the new company on tenuous ground. It could be denied local facilities or the right to expand its long-distance facilities without violating the initial decision. Although the FCC deferral of major questions raised by the MCI application allowed the same issues to be argued over and over again in subsequent hearings, it was beneficial to MCI to gain initial operating authority on narrow grounds rather than having to fight a more drawn out regulatory battle for a chance at a broader authorization.

Following the 1969 decision in favor of the MCI construction applications, the carriers petitioned for a rehearing but were refused. AT&T then filed suit in the appeals court attempting to overturn the decision. Meanwhile MCI filed modifications to its original construction applications to take account of changes in technology and its own plans during the years of regulatory consideration. This provided an occasion for further regulatory opposition by the existing carriers because the 1969 decision applied specifically to the 1963 construction applications. The carriers requested a new evidentiary hearing on the modifications to the construction permits, asserting that the more expensive equipment would be beyond MCI's financial capabilities and that the increased service that could result from the proposed equipment amounted to a different class of service than had been authorized by the commission. In January 1971 the commission denied the requests of the carriers to hold additional hearings and granted the modified construction permits to MCI.[19] In that same year, AT&T withdrew its appeal of the 1969 decision and the construction of the system was completed. It opened for service in January 1972.

It was over seven years from the initial filing of the MCI applications in late 1963 to the final approval of the applications in early 1971. The actual construction of the system required approximately seven months and was completed during the year 1971. The chairman of the board of MCI testified that MCI had spent $10 million in regulatory and legal costs in order to secure approval for its network.[20] The estimated cost for the actual facilities between St. Louis and Chicago was under $2 million.[21] The capital costs and construction times involved in the Chicago–St. Louis route indicate low barriers to entry relative to most manufacturing enterprises. However, the regulatory costs and delays indicate extremely high barriers to entry. In this case, MCI's application was approved at each stage from the initial hearing examiner's decision in 1967 through the final approval in 1971 yet the total delay involved was enormous.

Following the 1969 decision authorizing the MCI entry, a large number of additional construction applications were filed by various parties to provide similar kinds of service. Rather than holding separate proceedings on each set of applications, the FCC initiated a broad policy inquiry regarding the desirability of specialized common carrier competition and invited comments from interested parties. Over two hundred parties responded with the predictable split of opinion. The existing common carriers were against any grant of new applications and challenged the right of the commission to authorize extensive new competition. They asked for delays of various kinds, ranging from individual evidentiary hearings on each application to a hold on new authorizations until the MCI facilities were actually in operation and could be evaluated. The carriers claimed that the new competition would cause great disruption in the established communications system and lead to higher rates for customers, as well as causing frequency congestion and increased costs through duplication of facilities. The potential new carriers, their equipment suppliers, and their potential customers all supported new entry. In May 1971 the FCC announced a fundamental policy decision in favor of increased competition. The commission stated:

> We find that: there is a public need and demand for the proposed facilities and services and for new and diverse sources of supply, competition in the specialized communications field is reasonably feasible, there are grounds for a reasonable expectation that new entry will have some beneficial effects, and there is no reason to anticipate that new entry would have any adverse impact on service to the public by existing carriers such as to outweigh the considerations supporting new entry. We further find and conclude that a general policy in favor of the entry of new carriers in the specialized communications field would serve the public interest, convenience, and necessity.[22]

The broad policy statement by the FCC in favor of competition ensured that regulatory barriers would be at a minimum. Each carrier would not have to go through the extensive proceedings that were required of MCI in order to validate its right to enter the market. The FCC still retained the right to make a carrier-by-carrier determination when actually faced with applications, but relegated many of the policy issues to the background with its Specialized Common Carrier Decision. As individual applications came before the FCC for review, the established carriers continued to raise objections to them, but the objections were largely dismissed as relating to issues already decided in the Specialized Common Carrier decision.

The reduced regulatory barriers to entry induced several companies to establish microwave systems in competition with AT&T. MCI opened its Chicago to St.Louis link in January 1972 to inaugurate the era of new competition. MCI rapidly expanded its network, both through building its own facilities and through mergers with other fledgling microwave competitors. The company set itself up as a holding company with controlling interest in geographical operating companies. Its initial $2 million, two-city system was expanded to a coast-to-coast, $80 million, forty-city system by the end of 1973. Southern Pacific Communications Company, formed after the initial favorable MCI ruling as a wholly owned subsidiary of the Southern Pacific Company, also completed a coast-to-coast microwave system by the end of 1973. The parent company operated the country's largest private communications system and had extensive experience with microwave technology before the field was opened to commercial exploitation. The communications subsidiary used many of the sites of the parent company's private system but maintained separate systems for commercial and private use.

MCI and Southern Pacific established analog systems primarily designed for voice transmission. Computer data could be sent over an analog system by using a modem to translate between digital and analog signals at each end, but greater accuracy could be obtained with an all digital network. To meet the growing demand for transmission of computer data, Datran was established as a subsidiary of University Computing Corporation to build an all digital microwave network. Datran's network included some local distribution facilities and opened for service in 1973.[23]

With the 1969 AT&T private-line rates and free interconnection, new competitors would have quickly set up nationwide networks on the dense communications paths between major cities and tied into the AT&T network for local distribution and low-density feeder lines. The economies of scale of microwave systems protected AT&T from direct competition on low-density routes and made it profitable for the competitors to use AT&T facilities to extend their operations at the existing rates. If market power was to be maintained, a reduction in price or increase in barriers to entry was necessary. The least expensive investment in barriers to entry was legal expense to maintain interconnection limitations.)If the new competitors could be completely isolated from the AT&T system as in the early days of competition, there would be practically no market at all for their services. Even if full interconnection was eventually required, the ability to fight the interconnection issue before state regulatory agencies as well as the FCC provided opportunity for extensive delay.

A more expensive method of retarding competition would be to modify the rate structure to reduce the incentive to enter. While across-the-board rate reductions would be very expensive for the company relative to the new competitors, a restructuring that reduced prices on routes most open to competition and increased prices on more protected routes was a feasible protective device. Two kinds of restructuring were desirable to protect the company. The first was to increase rates on low-density routes protected from entry by economies of scale and decrease rates on high-density routes subject to entry. The second was to increase the price of short-haul lines and decrease the price of long-haul lines to reduce the ability of competitors to use short-distance AT&T lines to connect customers to the competitive network. The optimal strategy for maintaining market control would be to use interconnection restrictions to limit the potential scope of competitive networks as much as legally possible and then to use price modifications to reduce the competitive opportunities in the segments left open to competition.

Interconnection rights had been considered in the Specialized Common Carrier proceeding and the commission had concluded:

> We reaffirm the view expressed in the *Notice* (paragraph 67) that established carriers with exchange facilities should, upon request, permit interconnection or leased channel arrangements on reasonable terms and conditions to be negotiated with the new carriers, and also afford their customers the option of obtaining local distribution service under reasonable terms set forth in the tariff schedules of the local carrier. Moreover, as there stated, "where a carrier has monopoly control over essential facilities we will not condone any policy or practice whereby such carrier would discriminate in favor of an affiliated carrier or show favoritism among competitors."[24]

MCI and other specialized carriers interpreted the last clause of the preceding quote to mean they were entitled to service from Bell-controlled local companies on the terms offered to the AT&T Long Lines Department. They asserted that no further proceedings were necessary to establish their legal right to full interconnection.

AT&T denied the FCC's jurisdiction over interconnection and filed tariffs for local distribution of specialized carrier communications only with the various state regulatory commissions. The tariffs provided for local distribution of simple private-line service, in which one phone is connected only to one other phone, but not for the more complex types of private-line service such as FX and CCSA service. FX (foreign exchange) service is a version of private-line service in which the user ties into the local network of a distant city, effectively gaining a local number in the distant city, rather than only gaining the ability to communicate with a single location as in ordinary private-line service. CCSA (common control switching arrangements) allow a subscriber to link a system of private lines through telephone company switches to provide a private network. Because FX and CCSA accounted for a substantial portion of private-line demand, the restriction to simple private-line service imposed a serious constraint on the new competitors.

MCI challenged the AT&T action before the FCC. The chief of the FCC Common Carrier Bureau ruled in October 1973 that MCI was entitled to full interconnection rights. Bell appealed the decision to the entire commission and asserted that it had been denied an opportunity for a hearing on the need for interconnection. Bell claimed that the local distribution services were properly intrastate services and not subject to the jurisdiction of the commission, that in any case it had not been ordered to provide interconnect services by the commission and was entitled to a full hearing before such an order could be issued. The company declared that the commission's analogy between the services provided by local operating companies to Long Lines and the services the local operating companies should provide to MCI was incorrect because the local companies did not provide services to Long Lines but engaged in a joint venture, and that MCI would not be allowed to participate in that joint venture. In April 1974 the commission ruled against Bell and asserted that it had jurisdiction, that a proper hearing had been held, and that Bell was required to provide a full complement of interconnect services to the specialized carriers. The commission ordered that within ten days Bell was required to "furnish to the specialized carriers for their authorized interstate and foreign communications services, interconnection facilities similar to those presently provided to Bell's Long Lines Department on a non-discriminatory basis."[25]

In May 1974 Bell filed tariffs allowing interconnection. The new interconnection tariffs allowed the provision of FX and CCSA service by the specialized carriers, but restricted local service to points where the customer had "a regular and continuing requirement to originate or terminate communications" except in certain special cases. This restriction meant that a specialized carrier could not order a Bell line to terminate on the specialized carrier's premises and thus use the Bell facilities to complete service to customers outside the range of its own facilities. In July 1976 the FCC ruled that the restriction was unreasonable and in violation of previous FCC orders. With the July 1976 order, the specialized carriers obtained complete interconnection rights.[26]

The rights to interconnection were upheld by the FCC at every step, but the regulatory procedures used delayed full interconnection until five years after the Specialized Common Carrier decision in which entry and interconnection were affirmed. Bell's interconnection restrictions were an effective and inexpensive method of delaying the spread of specialized competitors. Because each new tariff created the opportunity for further hearings and the only penalty imposed if the tariff was found illegal was the requirement to file a new tariff, AT&T incurred little risk in resisting interconnection requirements. The resistance allowed AT&T to be certain of a delay in implementation of interconnection and to gain the possibility of a favorable FCC ruling or the exhaustion of the new competitors before the interconnection was implemented.

AT&T gradually modified its price structure to reduce entry incentives as the regulatory barriers to entry were reduced. Following the initial commission decision in favor of allowing MCI to compete in the specialized carrier segment in August 1969, AT&T filed a new private line tariff known as Series 11,000. The Series 11,000 offering instituted a special low-price service in certain specific areas of heavy demand, including the area proposed to be served by MCI. It was limited to two-point service and designated an experimental tariff to provide information on alternative tariff structures. The FCC stated that it could not determine whether or not the charges were "just and reasonable" but that no purpose would be served by a short suspension. The rates were allowed to go into effect as filed and the question of their lawfulness was combined into the interminable investigation of the private-line rate structure begun by the Telpak tariff eight years earlier.[27]

The Series 11,000 rates can be viewed as a signal to the potential new entrant that, after regulatory remedies were exhausted, market competition would be used. MCI could not compute the expected profitability of its competitive services based on the rates in effect at the time it entered the business. If a potential entrant faces significant entry costs but knows the existing firms will cut prices when actual entry is made, prices may be kept above the limit price level without attracting entry. In this case, MCI continued to prepare for actual entry in spite of the new prices.

When MCI actually began service in January 1972, there was considerable support within AT&T for a direct price response on the St. Louis-Chicago route in order to stop MCI expansion, but the company decided to defer pricing action for further study. The minutes of the Bell Presidents Conference in May 1972 reported:

> A number of the conferees expressed disappointment that AT&T had not filed an immediate competitive response to MCI on the Chicago–St. Louis route. An immediate response, it was pointed out, would have provided clear notice of our intention to compete. To delay is to risk the prospect that, once one or more specialized common carriers become going businesses, regulatory authorities might seek to assure their continued viability regardless of the economic justification for their survival.
>
> On the other hand, it was pointed out, that MCI does not—and may not for some time—pose a sufficient threat to warrant a basic change of policy without a thorough examination of the consequences, particularly the consequences to our message toll business, which accounts for some 87 per cent of our interstate revenues. Also, nationwide average pricing has served the nation well; isn't it worth a fight to prevent its further erosion, certainly before we breach the principle ourselves?[28]

At the time MCI had thirty-nine customers using a total of forty-eight circuits. AT&T Chairman John DeButts announced at the May 1972 conference:

> With respect to MCI, the EPC did in fact decide at one point that as soon after their filing as practicable the Bell System would respond

> with an "exception tariff" that would take account of the changed conditions in the Chicago–St. Louis route. The fact that on maturer reflection the EPC decided to defer this filing should not be taken as a sign of unreadiness to compete. The fact of the matter is that we realized we simply did not have the information on which to make an intelligent decision and in the absence of that information it didn't make sense to commit ourselves to a course of action, involving as it would a breach of so basic a principle as nationwide average pricing, from which there would be no turning back.
>
> In any case, the necessity [*sic*] studies are now well underway that will put us in a position to examine all the alternatives intelligently and decide—not whether we'll compete, that's decided—but how.[29]

As the specialized carriers built facilities during 1973, AT&T made its competitive response. In November 1973 the company filed a new rate plan for voice private lines. The country was divided up into regions of high-density lines and low-density lines. The new rates reduced the price to $.85/mile for lines between high-density centers and increased the rates to $2.50/mile for lines between low-density centers or between a high-density center and a low-density center. AT&T reported that the total effect on its revenue would be nil, because the decreases in rates for 3,000 customers would cancel out the increases in rates for 16,000 customers. Only the competitive private line rates were restructured by route density. The monopoly switched voice network rates remained independent of route density.

From AT&T's point of view, the rate structure changes were an effective method of reducing the incentive to enter without losing substantial profits. In 1965 the FCC had instituted its first formal investigation into the charges and rates of return on interstate services and had ruled in 1967 that interstate services as a whole should be limited to the general range of 7.0 to 7.5 percent on invested capital. The rate of return was not a rigid limit but a general guide. The commission ruled that rate adjustments would not automatically be required if earnings were outside the specified range, but that earnings outside the range would cause the commission to "consider what further action may be required in light of the then current conditions."[30] In contrast to the two-year proceeding to determine the overall rate of return, the consideration of the appropriate rate structure begun by the Telpak tariff was then in its twelfth year with no end in sight. Thus it was a reasonable assumption for AT&T that its overall rate of return was subject to regulatory constraints but that the charges for individual services were not. Consequently, a tariff which produced the same amount of revenue but restructured the rates to reduce entry was unlikely to be overruled by regulation. If it was challenged by the FCC, the only likely penalty was an order to replace it with a new tariff.

The tariff adjustments would not be totally costless to AT&T because the adjustments could be expected to alter the pattern of demand, possibly requiring new construction on routes with reduced prices and causing excess capacity on routes with increased prices. Because the private-line market on which the price adjustments were made was a small part of the total communications capacity, the capacity cost problems resulting from the tariff modifications could have been expected to be small.

Following complaints from the specialized competitors, the FCC suspended the tariff for the maximum ninety days allowed by law and began a proceeding to examine the reasonableness of the new tariff. An expedited hearing was held to minimize the potential adverse effects of a long continuation of the tariff if it was found to be illegal. After two years the tariff was found to be unreasonable and in violation of the Communications Act. The FCC ordered AT&T to file a replacement tariff.[31]

The new tariff was known as the Multi-Schedule Private Line (MPL) rate scheme. It was similar to the previous tariff in that it divided the country into centers served by low-capacity facilities and centers served by high-capacity facilities with lower rates for the high-capacity areas. Rather than the previous flat rate per mile once the density was established, the MPL tariff contained a decreasing charge per mile with increases in the line distance. The specialized carriers opposed the MPL tariff, claiming that it only accentuated the discriminatory aspects of the rejected tariff.[32] The high charges for short-haul lines reduced the ability of the specialized carriers to complete their systems with AT&T facilities.

The FCC suspended the tariff for the maximum three-month period and instituted an investigation into the MPL tariff. After the three-month suspension, the MPL tariff went into effect in August 1976 while the FCC investigation into its lawfulness continued. The investigation of MPL rates became very extensive. The first phase of the MPL investigation was the determination of proper costs to be assigned to the service. At the conclusion of that phase, the administrative law judge ruled in early 1979 that AT&T's cost allocation procedures were so flawed that the MPL tariff should be rejected rather than moving into additional stages of the investigation.[33] However, the inadequate cost figures also left the commission with little information on which to specify a tariff. Competitors of AT&T suggested that the last legal tariff (pre-1973 rates) should be restored until AT&T filed a new cost-justified tariff. AT&T challenged the legal authority of the commission to impose the old rates. It stated:

> There is no basis for such a Commission action which would be tantamount to an unlawful Commission prescription of rates under section 205(a) of the Communications Act—without the requisite findings properly supported on the basis of a full evidentiary record. Thus, the Commission may not prescribe these old rates for Bell's private line services without fully supported findings that such rates are just, reasonable, and otherwise lawful under now current conditions, which are vastly different from those existing at the time these old rates were in effect.[34]

The commission's September 1979 review of the administrative law judge's decision and the parties' comments on it brought it to the conclusion that "the MPL rates are unjust and unreasonable and that the support material is unreliable and the phase I record does not support a prescription of rates."[35] The commission thus agreed both with the administrative law judge that the existing rates were improper and with AT&T that the commission had inadequate information to prescribe proper rates. Consequently it canceled further hearings on MPL, allowed the tariff to remain in effect, and instituted a new and broad-ranging inquiry into the general structure of AT&T's private-line tariffs. The inquiry was concerned with the great complexity of the tariffs as well as the rate levels.

The MPL proceeding confirmed that the commission could not control individual service rates. So long as any tariff prescribed by the commission must be based on data that only AT&T can provide, AT&T can retain pricing freedom by only providing inadequate information. Although the commission can challenge an existing tariff, the challenges are meaningless if AT&T is allowed to maintain the tariff until it chooses to change it. The legal right of the commission to simply reject an illegal tariff and refuse to allow AT&T to offer the service until a legal tariff is filed has not been used because of the hardship it could impose on existing users. Commissioner Joseph Fogarty stated in his dissent to the MPL decision:

> I have been sorely tempted to move the Commission to cancel the existing MPL tariff as a remedy for AT&T's patent and persistent failure to justify its rates and to comply with statutory and rule provisions and outstanding Commission orders. Perhaps this is thinking the unthinkable, because cancellation of the AT&T MPL tariff would leave thousands of private line customers without essential service, and I want to protect these customers, of course. Nevertheless, it may be appropriate to think the unthinkable if the effect of ordering cancellation of an unjustified private line tariff on 30 days' notice would be the expedited filing by AT&T of a legally sufficient and reviewable tariff.[36]

MCI did not entirely reduce its rates to be competitive with the MPL rates. The final rate structure was complex, but an idea of the rates can be obtained from observing that at the beginning of 1978, the MCI marginal rates ranged from $.87 per month per mile for a voice grade circuit to $.54 per month per mile depending on the total volume ordered by the customer. At the same time, the AT&T marginal rates ranged from $4.40 per month per mile for short distances between low-density points down to $.40 per month per mile for long-distance connections between high-density cities. The lowest AT&T marginal rates were below the lowest MCI marginal rates and the highest AT&T marginal rates were above the highest MCI marginal rates. The choice between AT&T and MCI depended on the particular route and distance over which communication was needed.[37]

By 1975 the total telephone industry (including the new competitors) had revenues of $35,275 million, of which Bell received $29,581 million or 83.9 percent. Of the total revenues, interstate private-line service accounted for $1,068 million or 3.0 percent of the total telephone revenues. The specialized microwave carriers received revenues of $35 million or 3.3 percent of the private line market and 0.1 percent of the total telephone revenues.[38] Six years after the initial MCI authorization, the new competitors remained fringe firms with a tiny share of telephone revenues. The slow growth of competition after the MCI decision was in marked contrast to the rapid entry prior to 1900, which gave competitors approximately 37 percent of the total phones six years after the expiration of the original Bell patents.[39] Both cases involved new entry after a sharp drop in legal barriers to entry, but Bell was much quicker to compensate for the reduced market protection in the specialized carrier case than in the original telephone competition.

The only actual casualty of AT&T's protective measures was Datran, the company formed to provide digital transmission of computer data. Soon after Datran's application was approved, AT&T began building its own digital network for carrying data. AT&T's initial tariff for the digital system in 1974 provided for very low rates relative to the charges for similar communications capacity in analog mode. As a result of Datran's protest over the tariff, the FCC began a hearing on its validity while it remained in effect. In June 1976 the administrative law judge ruled that the data service rates were unjustified, predatory in intent, and discriminatory, but that there was inadequate information presented at the hearing on which to prescribe a new rate and that therefore the existing rates should continue in effect until AT&T filed a new tariff. The decision was appealed to the entire commission and essentially affirmed in January 1977. The commission concluded that the rates were being cross-subsidized by users of other AT&T services but that it also had too little information to enable it to prescribe a tariff. It ordered AT&T to file a new tariff.[40]

Between the intial decision and final decision, Datran declared bankruptcy. Service was suspended in September 1976, but the network was soon purchased by Southern Pacific Communications and the digital service was reestablished. AT&T made complex pricing modifications to its digital service and gradually expanded its network of digital data transmission facilities.

Competition in Switched Long-Distance Service

The initial competition was confined to private-line service. The main market for long-distance communication was ordinary dial-up telephone service. The fundamental difference between the two services is the length of time for which the facilities are contracted. Dial-up service allows the user to lease facilities by the minute, while private lines allow the user to lease facilities by the month. The technology and access arrangements are the same for private lines and dial-up facilities. Thus once the specialized carriers had long-distance communications capacity for private lines, they also had the capability to offer dial-up long-distance service. Just as with private lines, a dial-up call could proceed from the user's office to the local MCI office over local phone lines, then over MCI long-distance facilities to the MCI office in the destination city, and then over local phone lines from the MCI office to the destination phone.

The economic incentive to enter the switched long-distance market was strong. None of the pricing modifications which had reduced entry incentives in the private-line market had occurred in the switched market. In addition AT&T's switched long-distance prices included a component for the use of local facilities through the separations process while private-line calls were exempt from the separations process. The attractiveness of the switched long-distance market and the obstacles to expansion in the private-line market caused the specialized carriers to consider the possibility of expanding their operations beyond private-line service. No explicit monopoly of the switched long-distance market (referred to by AT&T as MTS for message telecommunications service) had been given to AT&T, but the original regulatory grants to the specialized common carriers did not explicitly authorize them to enter that market. Whether the new carriers had the right to offer MTS service simply by filing a tariff or whether explicit new authorization was necessary was an open legal question.

MCI asserted that it was not restricted in the scope of its service offerings. However, it also packaged its entry into the MTS market as a "shared private line" service in order to strengthen its position against challenges to its authority to offer the service. After filing a tariff, MCI began marketing the new service (called Execunet) in January 1975. With the Execunet service, a customer initiated a call by calling the local MCI office, and entering his customer number and the telephone number of the person to be called. The customer was charged by minutes of connect time and mileage to the distant city, plus a connection charge and a monthly minimum charge. MCI described the service as "shared FX" because it had the characteristics of foreign exchange private-line service except that the lines could be rented by the minute rather than by the month. From the customer's point of view, the Execunet service was ordinary switched long-distance service. The differences were that the rates were lower than Bell's and the service was less flexible because of the monthly minimum charge and the restricted nature of MCI's network relative to Bell's.

Soon after MCI began marketing Execunet service, AT&T subscribed to the service through an intermediary and decided that the service was competitive with its own long-distance service. AT&T then began discussions with FCC staff (including a demonstration of Execunet service for them) in an attempt to get Execunet classified as an illegal competitor to long-distance service. Following the informal discussions, AT&T filed a complaint on May 19, 1975, which the FCC staff referred to MCI for reply. During June MCI responded to AT&T's complaint by asserting that Execunet really was a private-line service. On July 2, 1975, six weeks after the original AT&T complaint, the FCC rejected the MCI tariff as being unlawful on its face because MCI was restricted to private-line service and Execunet was not a private-line service. The commission ordered MCI to discontinue the service within thirty days. The quick Execunet action without a formal hearing or investigation was in marked contrast to the drawn-out proceedings over disputed AT&T tariff provisions. The commission justified its summary action on the grounds that there were no real facts in dispute and that the original authorizations to MCI limited its scope of services to the point that no investigation was necessary to determine that Execunet was an unlawful service.

MCI appealed the commission's decision on procedural grounds. It contended that it had not been provided adequate notice and opportunity for hearing, and had not been given an opportunity to respond to AT&T's informal presentations prior to the filing of its complaint. The court granted a stay in the termination order and at the FCC's request remanded the issue to the commission for further consideration. The FCC then called for comments from interested parties on the right of MCI to offer Execunet. The controversy centered on whether the authorization of the new carriers to provide "specialized" service restricted them to private-line services. The Specialized Common Carrier Decision did not explicitly restrict the new carriers to private-line service, but some of the language of that decision could be construed as restricting the scope of competition. After receiving the comments, the commission reaffirmed its year earlier decision in June 1976 and ordered MCI to discontinue Execunet within thirty days of the order or at the time the stay was lifted by the appeals court.[41]

MCI again appealed the decision and in November 1976 the company obtained a partial stay of the commission order to discontinue Execunet service. MCI was allowed to continue serving existing Execunet customers but not to accept any new orders while the appeal was being considered. In July 1977 the appeals court reversed the FCC decision on Execunet. The court ruled that the commission had not defined the boundaries of the specialized communications field prior to the Execunet rejection and had not granted AT&T a monopoly over ordinary long-distance services. It ruled that the commission must find evidence that the public interest would be promoted by an AT&T monopoly in order to justify excluding competitors instead of allowing a monopoly simply because AT&T was there first. The court allowed MCI to offer Execunet to new customers over its existing facilities and remanded the case to the commission for further consideration.[42] A key legal ruling in the decision was that, because the FCC had not specifically limited the uses to which MCI's facilities could be put at the time construction permits were granted, the facilities could be used for any service for which MCI filed a valid tariff. Thus MCI was entitled to offer Execunet service on any of its existing facilities but not necessarily entitled to an extension of the service on new facilities that had not yet been constructed.

The FCC appealed the decision to the Supreme Court, but in January 1978 the Supreme Court declined to review it. Immediately after the Supreme Court's denial of review, AT&T announced that it would not provide local distribution facilities for Execunet service, justifying its refusal on the grounds that the previous FCC and court proceedings requiring connections applied only to private-line service. In February the commission upheld AT&T's contention that it had no obligation to provide local distribution facilities for Execunet service. MCI then returned to the appeals court and in April obtained another reversal of the FCC's ruling and an order that AT&T was required to provide local facilities. In May 1978 the Supreme Court denied AT&T's request for a stay in the appeals court ruling pending full consideration. AT&T and the FCC then filed formal appeals of the decision, but in November 1978 the Supreme Court declined to review the case.[43]

Immediately after the Supreme Court declined to stay the requirement that AT&T provide local exchange facilities for Execunet, AT&T filed a new tariff known as ENFIA (exchange network facilities for interstate access) to provide local service to MCI and requested permission to implement the tariff on one-day notice rather than the normal ninety-day notice. The ENFIA tariff provided drastically higher rates for local exchange lines provided to other companies to be connected to an Execunet type service than for local exchange lines provided for ordinary business purposes, and included both a fixed charge per month and a charge per minute of usage. In some cases, the charge for using the local exchange line would be higher than AT&T's long-distance charges, thus eliminating any incentive to use the MCI services. The justification for the higher charges was that existing procedures for dividing long-distance revenue between the local operating companies and AT&T's Long Lines credited some of the toll charges to local service. AT&T asserted that the local operating companies had a right to higher revenue for lines used to access the MCI network than for lines used only for local service because local service lines attached to AT&T's network would receive additional revenue from AT&T's toll calls.[44]

The specialized carriers objected to the ENFIA tariff as not properly cost justified, anticompetitive in intent, and inconsistent with court rulings in the Execunet cases. MCI attached an affidavit stating that in the fiscal year ending March 1978 the company had paid $3.8 million for local connections for its "shared private line service" (Execunet), but would have paid $11.6 million for the same services under ENFIA, converting the company's net operating income of $3.7 million for the year into a $4.1 million loss.[45] Rather than instituting formal hearings on the validity of the tariff, the FCC convened the various parties for negotiations in an attempt to work out a compromise.

After three months of negotiations, the parties signed an agreement in December 1978 incorporating a compromise on the ENFIA tariff. The agreement called for a charge of $19.83 per month per access line plus a charge per minute. The charge per minute was based on AT&T's allocation of local exchange costs to interstate toll use. The allocation procedure called for dividing interstate minutes (including ordinary long-distance and WATS but not private-line services) by total minutes of use, multiplying the resulting percentage by 3.29, and using the product as a percentage of local distribution costs to be charged to the toll services. The factor amounted to 5.5 cents per minute of interstate use at the time of the agreement. The agreement called for the specialized carriers to pay 35 percent of the 5.5 cents/minute factor on Execunet type calls so long as their combined revenues were under $110 million, 45 percent of the factor if combined revenues were between $110 million and $250 million, and 55 percent if combined revenues were between $250 million and $375 million. At the time of the agreement, Southern

Pacific Communications had added an Execunet type service and ITT Communications was considering the addition of an Execunet-type service. The agreement was made for a three-year period with the expectation that the FCC would rule on the proper arrangements before the agreement expired. At the time of the agreement, the FCC had begun a wide-ranging inquiry on the public interest questions involved in monopoly provision of long-distance service, an inquiry that was expected to last for five years, but the negotiators expected an interim ruling on local exchange provisions. MCI estimated that its monthly cost per local exchange connection for Execunet under the agreement would be $86.70 at the first step, $107.44 at the second, and $133.03 at the third.[46]

The ENFIA agreement at the end of 1978 established the new carriers' right to continue operating in the dial-up long-distance market while the FCC was considering the issue of whether or not a monopoly was necessary in that market. The high charges for local access lines effectively brought the new carriers under the separations process and eliminated the incentive for entry purely from differential separations treatment. However, the charges did not remove the incentive to enter from AT&T prices above the level of least costs. Rather than instituting a major restructuring of the MTS rates or a general rate reduction, AT&T allowed the carriers to continue entry. During 1979 the percentage rate of growth in competitive dial-up services was high, although the absolute amount remained an insignificant percentage of AT&T's services. The number of local lines used for accessing competitive MTS services rose from 15,000 in April 1979 to 20,000 in August 1979 with extensive expansion planned by the specialized carriers. In the year ending June 30, 1979, total revenues of the specialized carriers were approximately $50 million, well under the $110 million point at which the ENFIA rates were scheduled to rise. MCI was by far the dominant firm among the specialized carriers with almost $43 million of the $50 million total for the year.[47]

The court decisions requiring the FCC to allow MCI to operate in the MTS market until an explicit policy on market structure was developed essentially predetermined the result of the FCC inquiry. Because the inquiry is expected to last many years, extensive investments will have been made in facilities to offer the new services before a decision is reached. In addition, large numbers of customers will gain an interest in the continuation of the service. It is therefore practically certain that the

FCC will not order the specialized carriers to abandon their service and return the market to a monopoly when and if a final decision is reached in the inquiry. The actual existence of firms other than AT&T limits the FCC's options to approval of competition or the establishment of a cartel with specific market-sharing arrangements.

The combination of usable facilities in place from another service and the absence of an explicit regulatory policy produced the anomalous result of entry into the dial-up long-distance market while the regulatory commission maintained that the market should be a monopoly. In this case, the regulatory lag that normally favors the established firm against all changes favored the new entrants. If the FCC had been able to hold a legally defensible hearing prior to the implementation of the Execunet service, it would almost certainly have ruled against the MCI entry into the MTS market. Because the commission had not previously held hearings on the matter and could not do so quickly, the court allowed MCI to proceed with its service and granted it the necessary interconnection rights to do so. The economic effect of the "accidental" entry into the MTS market is far greater than that of the slowly won rights to compete in the private-line market. The court decisions moved competition from a specialized segment of long distance communications into the entire market.

The size of the MTS market makes economies of scale less of a barrier to entry than in the private-line market, because more routes can efficiently use multiple microwave systems for combined MTS–private-line communications than for private-line communications alone. Consequently, if reasonably equal local access charges are established for AT&T's Long Lines department and the specialized carriers, and if no further regulatory barriers to competition are imposed, the charges for long-distance calls can be expected to decline toward the level of actual costs plus the separations local subsidy. Some AT&T pricing structure modifications can be expected because of the differential barriers to entry on high-density and low-density routes. However, the magnitude of the necessary price cuts to deter competition on high-density routes would be much greater in the MTS case than the private-line case. Thus it may not be in AT&T's interests to deter entry through pricing action at least until the competitors have a much greater share of the market than they had at the end of 1979.

Evaluation of Long-Distance Competition

The period 1956–1979 was a time of drastic reduction in barriers to entry in the long-distance market with minor changes in AT&T's market share. At the end of the period the new competitors as a whole had an insignificant percentage of the total long-distance revenues, but the market share of the independents was increasing and promised to become substantial unless deterrent action was taken by AT&T. The reduction in regulatory barriers to entry in the private-line market had induced important changes in the structure of private-line rates as AT&T compensated for reduced regulatory protection by reducing the incentive to enter.

The regulatory barriers were reduced in a gradual and experimental manner rather than as a structured plan to bring competition to the industry. The granting of private microwave licenses in 1959 was a minor event in itself in terms of both regulatory policy and economic significance. It was primarily the result of new information showing adequate availability of frequencies. The AT&T rate reductions via the Telpak tariff for large users eliminated the economic incentive for private systems and prevented the widespread use of the new authority. Yet in retrospect, the 1959 decision to grant private microwave licenses was the key event in the gradual opening of the long-distance industry to competition.

The original MCI application was predicated on the policy of open entry for private users. Because private users had the right to build a system for their own use, it was a small step to allow an intermediary to build a system for several users to share. The original application was for a tiny system designed for specialized needs. Thus it was easy for the commission to approve it without dealing with the broader questions of competition in the industry. It was obvious that the original MCI system could not have any detrimental impact on the established communications systems because of its tiny scope, and it could potentially benefit some users. It also seems likely that the approval of the MCI application was partially the result of commission dissatisfaction with AT&T's reaction to the private microwave authorization. AT&T had unilaterally canceled the impact of the authorization for private microwave with its discriminatory Telpak tariff. Although the commission was unwilling or unable to formally cancel the Telpak tariff, it expressed concern about the price discrimination embodied in the tariff. The blatant discrimination in favor of large users weakened AT&T's argument that the company used its nationwide monopoly with averaged rates to subsidize small-scale and high-cost users with earnings from high-profit users. Thus the Telpak tariff may have encouraged the FCC to approve the MCI application which promised to reduce the price for users too small to qualify for Telpak.

With the approval of the MCI application, precedent was established for the approval of the other specialized carrier applications. The various interconnection orders were natural outgrowths of the approval of specialized carriers in order to make the services useful and to fulfill the long-established common carrier obligations of the local carriers to provide service to all who request it. The biggest expansion of competition, into the switched long-distance market, was established against the wishes of the FCC by the appeals court. The authorization was based on a technical legal issue rather than a determination that free entry was the optimal public policy. Because the original construction approval had not limited the use of the facilities, the FCC was not entitled to limit the scope of service after the facilities were built. The FCC was entitled to grant AT&T a monopoly if it followed appropriate hearing procedures before doing so, but it could not assume a monopoly was the appropriate public policy without formally evaluating the question.

AT&T's actions were consistent with what would be expected of a profit-maximizing dominant firm. The general strategy used was to delay entry as long as possible through regulatory hearings, to restrict the scope of actual entry as much as possible through interconnection restrictions, and as a last resort to reduce the incentive for entry through price cuts on the services subject to entry. Regulatory delay and interconnection restrictions were the least expensive method of deterring entry. Price cuts were feasible so long as they could be confined to narrow classes of customers such as users of over sixty private lines or private-line users on high-density routes. So long as the new competitors retain a small share of the market, broad-based price cuts to reduce the penetration rate are unlikely to be profitable.

The FCC's role in establishing the competition was to remove regulatory barriers to entry rather than to positively promote competition. The only entry promoting activity of the FCC was the requirement for full interconnection. Yet that requirement had been established by the independent telephone companies outside of the regulatory context in 1913 as an outgrowth of antitrust pressure. Thus it is likely that without regulation, interconnection would have occurred either voluntarily or as the result of an antitrust suit. Other than ordering interconnection, the role of the FCC was to delay entry from what would have occurred without regulation and to impose significant legal costs and uncertainty on potential entrants. No real control was exercised over AT&T's price structure. Although various tariffs were declared illegal, the commission left it up to AT&T to propose a new tariff each time with no safeguard that the new tariff would be any more satisfactory than the previous one. Assuming interconnection would have been required in the absence of regulation, entry would have been more rapid if the interstate market had been completely free from regulation than it was under the FCC's procompetition policies.

9 New Competition II: Terminal Equipment

The Terminal Equipment Market in 1956

Terminal equipment is the equipment that terminates a telephone wire on the customer's premises. It is also known as customer premise equipment. The most common kind of terminal equipment is the ordinary telephone set. Other common varieties of terminal equipment include key telephone sets (KTS), private branch exchanges (PBX), and modems. A KTS is used by businesses with a small number of telephone lines and several telephone sets. Each line is connected to each set and the user selects a line by pushing the appropriate key. A PBX is used by larger businesses with a substantial number of telephone lines and many telephone sets. The lines from the telephone company do not connect with each set but end at a customer switching center. The switching center is connected with each telephone set and connects the set to the appropriate outside line when the set is being used. A modem (modulator-demodulator) connects a telephone line to a computer or computer terminal. The modem translates signals from analog mode (used in telephone lines) to digital mode (used in computers). Terminal equipment is used to translate between the electrical signals on a telephone wire and the external form of the signals and to provide signals to control the communication path. In the ordinary telephone set, the dial pulses signal the central office to establish the correct communications path and the handset translates between voice and electrical signals.

In 1956 the market shares for terminal equipment at the final user level were exactly the same as the market shares for local telephone service. Bell was by far the dominant firm with approximately 85 percent of the market with the remainder divided among a large number of independent telephone companies. General Telephone was the largest of the independents with 5 percent of the market. Each telephone company provided the terminal equipment to its customers. Manufacturers of terminal equipment sold only to telephone companies, not to final users. Because Bell operating companies purchased essentially all of their terminal equipment from Western Electric, Western Electric's share of the manufacturing market was approximately the same as Bell's share of the final user market. Before 1955 the independent telephone companies purchased their terminal equipment from a number of equipment manufacturers of which Automatic Electric was the largest. General Telephone's acquisition of Automatic Electric in that year created a vertically integrated structure for the largest independent. Later mergers of additional independent telephone companies with equipment manufacturers increased the tie between equipment manufacturing and telephone service but left a small portion of the market for unaffiliated equipment manufacturers.[1]

Barriers to entry into the final user market were prohibitive. All telephone companies required that only equipment provided by them could be attached to telephone lines provided by them. A customer could not legally purchase a telephone set and attach it to a telephone company wire. Some unauthorized attachments occurred but there was no organized retail market in telephone sets. The tie between telephone service and telephone sets meant that the only method of entry into final user terminal equipment was to establish a telephone company. State regulatory authorities generally prohibited competitive telephone companies and substantial economic barriers to entry in local telephone service also existed, making entry into telephone service an unattractive route for entry into terminal equipment.

At the manufacturing level, the supply-side barriers to entry were quite low. Patent control was insignificant, especially after the licensing provisions of the 1956 AT&T Consent Decree. Economies of scale were no greater than in many electrical appliances. Many companies had the technological capability to manufacture terminal equipment. The demand-side barriers to entry were much higher. The small size of the market for telephone companies with no affiliated manufacturer made economies of scale a potential barrier to entry into that submarket. The integrated companies would occasionally purchase from suppliers other than their own subsidiary, but only to meet excess demand or to acquire a product which the subsidiary did not manufacture. Sale of a standard item, such as an ordinary telephone set, to an integrated company was essentially impossible, but specialty items for small submarkets could be sold to the integrated companies for remarketing to their customers.

The limited barriers to entry in the manufacture of terminal equipment and the large total size of the market implied that a substantial competitive market in terminal equipment would exist if the tie between telephone service and terminal equipment could be broken. In unregulated markets such tying contracts were a violation of the antitrust laws.[2] However, regulation protected the telephone companies from a straightforward antitrust challenge. Thus entry into the final user market required a regulatory challenge to the tie between terminal equipment and telephone service. The telephone companies' ability to withstand regulatory scrutiny of the tie was increased by the long-established character of the practice. From the very beginning, Bell had prohibited the attachment of any non-Bell-owned equipment to its lines. Independent companies also adopted that practice. Thus the maintenance of the status quo at the time regulation was imposed included maintenance of a tie between telephone service and terminal equipment. The telephone companies asserted that the tie was necessary in order to protect the network from harm due to malfunctioning equipment. To gain regulatory disallowance of the tie, a potential entrant faced the formidable task of disproving the assertion of network harm, while most of the relevant information was controlled by the telephone companies.

The tie extended the telephone companies' market power from telephone service to the terminal equipment market. The extension benefited the companies through regulatory considerations, managerial simplification, an increase in total barriers to entry, and the ability to practice price discrimination. If the return on capital is limited but above the cost of capital, the telephone company will benefit from monopolizing terminal equipment because it increases its rate base and consequently its allowable level of profits. Monopolization of terminal equipment simplifies managerial tasks and reduces the risk to the company. Standards for terminal equipment are determined internally rather than by an independent body or by negotiation with terminal suppliers. Customer demand will not fluctuate with new products introduced by competitors. Products will not be made obsolete by competitors. The managers can develop long-term plans without predicting the actions of competitors.

Total barriers to entry can be increased by the tie between the two products by forcing an entrant to enter both at once. So long as entry into telephone service is absolutely blocked, this is of no importance. But if legal barriers to entry should be withdrawn, then the economic barrier to entering both markets is greater than the barrier to either one alone. If a competitive market existed in terminal equipment and a large number of customers owned equipment that could be used with any telephone system, the task of setting up a competitive telephone system would be easier than if all telephone equipment was tied to the telephone network.

Price discrimination can be achieved through the tie because customers use variable proportions of telephone lines and terminal equipment. In general, the more intensive user of the telephone lines will have the greater number of telephone sets per line. The residential user with a single line but multiple extension sets generally values telephone service more highly than the user with only one set. Consequently maximum revenue can be obtained by setting a relatively low price on the basic service and a relatively high price on terminal equipment beyond the minimum necessary for operation. Because such a scheme places the highest price-cost margin on the item for which entry is the easiest, a formal tie is a necessary ingredient.

Incentives to enter the terminal equipment market at the final user level came from two sources. First, telephone company provided terminal equipment was priced above competitive costs. Second, the telephone companies were slow to respond to specialized needs. Essentially all residential customers in 1956 used a plain black telephone. Business customers had slightly greater variety but were still limited to a few standardized types of equipment. Little attempt was made to seek out and satisfy specialized desires for either style or function.[3] In more recent years, opportunities for innovative functions through advanced electronics technology have provided further incentive for entry into terminal equipment.

The Hush-A-Phone and Carterphone Decisions

The first significant challenge to the carriers' ability to maintain absolute control over attachments to the telephone system came in the Hush-A-Phone case. The Hush-A-Phone was a cup-like device that snapped on to the telephone instrument to provide speaking privacy and shield out surrounding noises. It was a passive nonelectrical device that directed the speaker's voice into the instrument. After the device had been in use for a number of years, AT&T began informing distributors and users of the Hush-A-Phone that the device violated the tariff forbidding the attachment to the telephone of any device not furnished by the telephone company.

In December 1948 Hush-A-Phone filed a complaint with the Federal Communications Commission and requested permission to sell the Hush-A-Phone without telephone company interference. The commission held hearings and oral argument on the complaint and in December 1955 ruled against Hush-A-Phone. The FCC found that the device did not impair the facilities of the telephone companies, but that it did reduce the quality of the conversation on the actual line on which it was used because it could reduce the volume and distort the voice of the person speaking into it. The commission disregarded the benefits to intelligibility from filtering out background noise that the Hush-A-Phone produced because the telephone companies could provide different devices that accomplished the same effect.

Hush-A-Phone Corporation did not accept its loss in the seven-year FCC proceeding as final and appealed the decision to the appeals court where it won a clear victory and set an important precedent on the limitations of the telephone companies to restrict private use of the telephone. The crucial issue was that the only harms found by the FCC were private to the parties conversing with the aid of the device and had no effect on the other subscribers. The court denied the telephone company's right to restrict private usage if no public harm resulted. The decision stated in part:

> The question, in the final analysis, is whether the Commission possesses enough control over the subscriber's use of his telephone to authorize the telephone company to prevent him from conversing in comparatively low and distorted tones. It would seem that, although the Commission has no such control in general, there is asserted a

> right to prevent the subscriber from achieving such tones by the aid of a device other than his own body. Thus, intervenors do not challenge the subscriber's right to seek privacy. They say only that he should achieve it by cupping his hand between the transmitter and his mouth and speaking in a low voice into this makeshift muffler. This substitute, we note, is not less likely to impair intelligibility that the Hush-A-Phone itself, for the Commission has found that "whenever an enclosure is placed around the mouth of a person an intensification of frequencies below approximately 500 cycles occurs, and if the intensification is too great, a distortion or blasting effect results in the transmitter." . . .To say that a telephone subscriber may produce the result in question by cupping his hand and speaking into it, but may not do so by using a device which leaves his hand free to write or do whatever else he wishes, is neither just or reasonable. The intervenor's tariffs, under the Commission's decision, are in unwarranted interference with the telephone subscriber's right reasonably to use his telephone in ways which are privately beneficial without being publicly detrimental.[4]

The Hush-A-Phone case established the very significant principle that some public harm must be shown in order to justify restrictive tariff provisions and that it is not the right of the telephone company to regulate how a subscriber uses his telephone so long as the only impact of the subscriber's use is to affect his own conversations.

Following the November 1956 court decision, the FCC adopted an order requiring AT&T and its associated companies to file new tariffs "rescinding and canceling any tariff regulations to the extent that they prohibit a customer from using, in connection with interstate or foreign telephone service, the Hush-A-Phone device or any other device which does not injure defendants' employees, facilities, the public in its use of defendants' services, or impair the operation of the telephone system."[5] Prior to the Hush-A-Phone decision, the AT&T restrictive tariff provision had read: "No equipment, apparatus, circuit or device not furnished by the telephone company shall be attached to or connected with the facilities furnished by the telephone company, whether physically, by induction or otherwise."[6] Rather than deleting the provision, the company added an additional paragraph that said the restriction would not be construed to prohibit a customer from using a device that served his convenience so long as the devices did not injure the telephone system, involve direct electrical connection to the system, provide a recording device on the line, or connect the telephone company line with any other communications device. The new provision clearly allowed the use of the Hush-A-Phone but did not allow many other devices that could have come under the requirements imposed by the appeals court to allow devices that provided private benefit without producing public harm. The commission accepted the new AT&T tariff without specifically approving it or ruling on whether or not it complied with the court imposed requirements.

Soon after the revised tariff went into effect, Carter Electronics Corporation began marketing a device called a Carterphone that connected mobile radio telephone systems to the telephone network. The Carterphone contained a cradle into which an ordinary telephone handset could be placed. The Carterphone transmitted voice signals from the mobile radio transmitter into the telephone handset and converted the voice signals received from the handset into radio signals for broadcast to the mobile radio-telephone without the need for a direct electrical connection between the two. The device violated the AT&T tariff because it provided a connection between telephone lines and other channels of communication, but it at least arguably fit within the Hush-A-Phone requirements because it was merely receiving voice signals from the telephone set and transmitting voice signals back to the telephone set. In response to a Carter inquiry, the commission informed the company in 1960 that the Carterphone did not violate any commission rules but that it appeared to violate the AT&T tariff. Carter continued to produce and market the device. AT&T asserted that the tariff prohibited the Carterphone and threatened to suspend telephone service to customers who used the Carterphone.

Carter again sought FCC assistance and was informed that the device appeared to violate the tariff and that the commission could only change a tariff after a hearing. Rather than requesting a formal hearing before the FCC, Carter filed an antitrust suit against AT&T. The court then passed the matter back to the FCC for action rather than proceeding directly with the antitrust suit. In 1966 the commission instituted an investigation into the Carterphone, including its effect on the telephone system, whether it violated the tariff, and whether the tariff itself was lawful. The telephone companies suggested a variety of methods in which the Carterphone could cause harm to the network, but were not able to specify actual problems that had arisen with the Carterphones in use. The hearing examiner found that the harms were speculative and of a nature that could be caused by the human voice as well as the Carterphone. He concluded:[7]

> In general, there is no reason to anticipate that the Carterphone will have an adverse effect on the telephone system or any part thereof. It takes nothing from the system other than the inductive force of the electrical field in the earpiece of the handset, which force is dissipated into the atmosphere in any event. It puts nothing into the system except the sound of a human voice into the mouthpiece of the handset, and that is the precise purpose for which that portion of the system is engineered.

In June 1968 the commission determined that the Carterphone violated the tariff but that the tariff itself was illegal and violated the requirements of the court and the commission in the Hush-A-Phone case. The commission ruled that the tariff "has been unreasonable and unreasonably discriminatory since its inception" and ordered the companies to file new tariffs to allow all devices that did not cause actual harm. The carriers were given permission to include restrictions on harmful devices and to specify technical standards to be met before a device was connected to the network.[8]

AT&T's argument for excluding the Carterphone and other devices from connection with the telephone network was based on the assertion that total control was necessary in order to prevent improper equipment from causing malfunctions in the telephone network. The arguments were similar to those of IBM in the 1936 case examining the validity of IBM's tie between tabulating cards and tabulating machines. The Supreme Court ruled in that case:[9]

> Appellant [IBM] places great emphasis on the admitted fact that it is essential to the successful performance of the leased machines that the cards used in them conform, with relatively minute tolerances, to specifications as to size, thickness and freedom from defects which would affect adversely the electrical circuits indispensable to the proper operation of the machines . . .
>
> Appellant is not prevented from proclaiming the virtues of its own cards or warning against the danger of using, in its machines, cards which do not conform to the necessary specifications, or even from making its lease conditional upon the use of cards which conform to them.

Although the Supreme Court found the tie-in between tabulating cards and tabulating machines illegal, it protected IBM's right to establish specifications for cards in order to avoid damage to its machines. The same principle was applied in later antitrust cases involving tie-ins. Thus the FCC's Carterphone decision that AT&T could not prohibit connections but could establish required standards to be met by connecting devices was an application of a long-established antitrust principle to the telecommunications industry.

Terminal Competition with Connecting Arrangements

In late 1968 AT&T filed new tariffs in response to the Carterphone decision. Rather than specifying technical standards to be met by interconnected devices as suggested by the commission, the company specified that competitive equipment could only be connected through a telephone-company-supplied device called a data access arrangement (DAA) if it connected a computer terminal to the network and a connecting arrangement (CA) if it connected a telephone system to the network. The telephone-company-supplied device would provide the network control signaling and would provide circuitry to protect the network against malfunctions in the customer-provided equipment. Many companies filed comments on the new tariff charging that it did not fulfill the Carterphone requirements, because it prohibited customers from providing signaling devices without showing that the devices could provide harm to the network. The interveners asked that the tariff be rejected outright or set for a formal hearing. The commission instead chose to allow it to go into effect without specifically approving it or ordering hearings on it.[10]

A large number of different kinds of CAs and DAAs were offered by the telephone company for specific kinds of connections. Prices varied with type of device and local telephone company tariff but ranged from nominal levels to monthly charges equal to the cost of the telephone service. The charges for the protective device eliminated the price incentive to enter the market for residential or single-line business phones. Entry was feasible into private branch exchanges (PBX), key telephone systems (KTS), and modems. The connecting device charges limited the potential market of new competitors to systems of about ten telephone sets or larger.[11] The requirement to secure a CA or DAA from the telephone company in order to attach competitive equipment continued the tie between telephone service and terminal equipment for a large part of the terminal market and placed competitors at a cost disadvantage to the telephone companies in the segments of the market in which competition was feasible.

The opening of the PBX and KTS market to competition caught AT&T unprepared for the new climate. Minutes of a late-1970 AT&T Presidents Conference made the following analysis of the company's position:

> Market programs are designed to service the selected accounts, handle demand and, when resources permit, initiate sales effort with those customers where the highest potential exists for increasing revenue. This leaves a large part of the market where contracts are infrequent and the communications consultant is not familiar with the specific customer. Our programs are not designed to provide the manpower to "service" and meet competition in the entire business market . . .
>
> The 701–740 vehicles in use are not competitive (features, floor space, appearance) with new cabinetized equipment. Customers are unwilling to use valuable space for bulky equipment . . .
>
> The development interval between inception and availability for service is excessive.
>
> The interval between the customer order and service is often unacceptable.[12]

By the following year, progress had been made in developing new and more competitive equipment, but there was also concern that the new equipment could lead to rapid replacement of the existing base. Material prepared for the late 1971 Presidents Conference reported:

> Our entire approach to the PBX market, from product inception to service offering, must become more market sensitive.
>
> The Western Electric Plant at Denver, Colorado, incorporates both the design efforts of Bell Labs and the manufacturing efforts of Western Electric under one roof, with a primary objective of decreasing design and manufacturing time. Two of the PBX systems manufactured at Denver, the 770A PBX and the 805A PBX, will play an important part in our future rate plans . . .
>
> A Hotel-Motel Service Package Offering is currently being developed utilizing the 770A PBX as the serving vehicle. The offering must

> maintain a proper relationship with our Feature-priced Tariff offering of the 701B PBX. We must consider that a package offering with a low monthly rate could result in numerous changouts of existing 701B installations to the new package offering. Some deterrent, possibly in the form of installation charges for the package system, must be introduced to discourage changeouts. Also, we must develop the package rate as low as possible in order to meet competition.[13]

The next year, the AT&T executives were more optimistic about their ability to meet the competition. The minutes of the 1972 Presidents Conference reported:

> There appeared to be general agreement among the conferees that we are well on the way to developing a fully competitive "product line" in the PBX/Centrex field. The "Denver concept" appears to be working well. Except at more than 400 lines we are abreast of the market—or ahead.[14]

Although the protective device charges and new AT&T products decreased the profits to be made by new competitors, the potential profits were still high enough to attract many companies into the market. As can be seen from Table 2, competitive replacement of key telephone systems and private branch exchanges increased at a high percentage rate during the first five years after the Carterphone decision. By 1974 the new competitors had installed enough telephone systems to produce the

TABLE 2. Bell and competitive PBX and KTS systems[a]

Year	Bell PBX + KTS ($ millions)	Interconnect PBX + KTS ($ millions)	Interconnect share (percent)
1969	1,486	0.4	0.03
1970	1,644	5.7	0.4
1971	1,775	18.0	1.0
1972	2,015	38.9	1.9
1973	2,227	67.6	2.9
1974	2,532	96.7	3.7

Source: FCC Docket 20,003, Bell Exhibit 20 (April 21, 1975).

a. All figures are based on systems within Bell operating territories. Dollar figures are converted to Bell System annual revenue basis for comparability between interconnect figures (often sold) and Bell figures (rented).

equivalent of $96.7 million annual revenue on a rental basis for a 3.7 percent share of the PBX and KTS market in Bell System territories. The interconnect business was divided among a large number of companies with no single company controlling the new competition. Much of the initial equipment was supplied by established foreign telecommunications companies but later equipment was designed by American companies for the interconnect market. One very significant potential competitor decided not to enter during the early period. IBM developed a PBX for the European market but decided against introducing it in the United States.

The advent of competition did not reduce AT&T's revenues from PBX-KTS systems or even reduce the historical growth rate of those revenues. In the five years preceding competition, AT&T's annual PBX-KTS revenues increased from $891 million to $1,486 million for a compounded growth of 10.2 percent per year. In the first five years of competition, AT&T's annual PBX-KTS revenues increased from $1,486 million to $2,532 million for a compounded growth rate of 10.7 percent per year. The overall market (in Bell territories) increased at an 11.4 percent per year compounded rate during the first five years of competition.[15] Thus all of the revenues of the new competitors fit into the additional growth of the market after competition, leaving AT&T revenues on their historical growth trend.

The PBX-KTS market size was increased after competition through price reductions and new products. New products incorporating greater use of electronics technology and improved convenience features were particularly significant in stimulating demand. By the end of 1973, customers had a choice of thirty-nine KTS systems (three from Western Electric) manufactured by fifteen different companies. The choice in PBX systems was even wider with 163 models available (thirteen from Western Electric) from twenty-eight different manufacturers. AT&T reorganized part of its development procedure in order to reduce the time between the design of new PBX equipment and actual delivery of the equipment to customers. The company's listing of its innovations in the PBX-KTS market indicated a greater number of innovations after the competitive era than before.[16] The large number of new features available from AT&T as well as more aggressive marketing of the devices by both AT&T and the competitors increased the size of the market.

The continued growth in Bell revenues from PBX and KTS systems of course did not indicate that the coming of competition had no effect on profitability in those segments. Total profit was decreased by reduced profit margins and through an increased rate of substitution of new products for obsolescent products. A depreciation study by South Central Bell for the state of Tennessee showed a rapid decrease in the expected life of large PBX equipment from almost twenty years before competition to just over ten years after competition. The decreased life expectancy came partially from customers switching directly to competitive accounts, but more significantly from customers switching to new Bell offerings designed to meet the competition. The Bell depreciation study stated:

> Most competitive cases result in the customer staying with Bell equipment. It is stated policy that the company will compete. The major tool of the company at this time to meet the competition is the 770A modular PBX . . . During 1973, over 50 customers had their existing PBX's replaced with the 770A, at their request. This action resulted in the retirement of over $945,000 from the PBX account . . .
>
> A new PBX system, the CSS 201 (customer switching system), will be introduced in 1975 and put into general use in the years following . . . It is expected that the CSS 201 will have considerable effect on retirements from all types of our present serving equipment. With the demand from our customers for smaller, quieter, more flexible systems, the CSS 201 will play a large role as our tool to meet competition. In meeting the competition, either changing present systems to the CSS 201 or removing the equipment where accounts are lost, retirements equal to or higher than present levels will be generated.[17]

Changing a large telephone system involves considerable effort in actual equipment removal and replacement (including extensive wiring) as well as in customer disruption. Consequently it is generally undertaken only with evidence of significant long-term benefit and careful preparation. Because of the expense of installation, the customer generally makes a long-term commitment to the new equipment. It is not feasible to switch to a new system temporarily while awaiting a new model from the telephone company or in order to increase bargaining power with the telephone company. Consequently only a small percentage of PBX users can be considered active participants in the market for improved telephone systems at any one time. This fact makes it cheaper for a telephone company to compete by offering new models of equipment to customers who might be lost to competition than by offering a general price reduction on existing equipment. A general price reduction to bring prices into line with competitive offerings would reduce revenue from all subscribers, not just those actively considering a competitive system. Maintaining old prices for old equipment and new lower prices for new equipment confines the benefits to the customers (and the cost to the company) to customers willing to pay the cost of switching systems.[18]

Certification and Direct Connection

After the institution of the post Carterphone tariffs, the commission began informally attempting to determine the need for the connecting devices through conferences with interested persons and a consulting contract with the National Academy of Sciences (NAS). The NAS concluded that network protection could be achieved by a program of standards for connected equipment and certification of equipment to ensure that the standards were met. Those results (as well as pressure from many customers and suppliers of interconnect equipment) induced the commission to begin a formal investigation in June 1972 to explore the technical feasibility of direct connection of customer-supplied devices without using telephone company connecting devices. At the same time, the commission convened a joint board of state commissioners and federal commissioners to consider the overlapping jurisdictional problems of requiring access to the network through customer-supplied equipment when that access involved both intrastate and interstate service.[19] Early in the proceeding, proposed standards were issued by an FCC advisory committee and much of the remainder of the proceeding was directed to modifications of the initial standards rather than the question of the feasibility of standards as such.

In October 1975, after three years of proceedings, the FCC adopted a registration and certification program to replace the carrier protective connecting devices. The commission concluded that the carrier required protective connecting devices were unnecessarily restrictive on the customer's right to use the equipment and were an unjust and unreasonable discrimination among users and among suppliers of terminal equipment. The commission prescribed that terminal equipment should be connected through standard plugs and jacks rather than direct wiring and that all terminal equipment (including that manufactured by the telephone companies) would be required to meet specified technical criteria in order to prevent harm to the network. The carriers vigorously protested the requirement that their own equipment would have to meet the certification tests and claimed that it would impose unnecessary expense, but the commission upheld the requirement as necessary to avoid competitive misuse of the technical standards.[20]

The certification program of October 1975 excluded key telephone systems, PBXs, main station telephones, coin telephones, and equipment connected to party lines because the commission believed the public had not had adequate opportunity to comment on the inclusion of major classes of telephone equipment in the registration program. However, after further consideration, the commission extended the registration program to PBXs, key telephone equipment, and main telephones in March 1976. A "grandfather" provision was also established that allowed equipment installed before the effective date of the program in accordance with the tariffs to continue to be connected without meeting the registration and certification requirements. The grandfather provisions applied both to the actual equipment installed and other equipment of that same model, allowing the telephone companies to continue to manufacture existing models without seeking certification.[21]

The telephone companies appealed the certification program to the appeals court and succeeded in obtaining a stay in implementation while the issue was being litigated. After one year (March 1977), the appeals court ruled in favor of the FCC but continued the stay in order to allow the companies opportunity to seek Supreme Court review. Following the refusal of the Supreme Court to review the decision in October 1977, the stay was lifted and the program became effective. The FCC refused Bell's request to delay the program until the completion of a separate proceeding on the economic impact of new competition. Following the refusal of the courts to stop the registration program and the refusal of the FCC to delay it any further, AT&T made one last effort at protecting the tie between telephone service and terminal equipment by proposing the "primary instrument concept." Under AT&T's primary instrument concept, the customer would be required to take at least one telephone set from the telephone company for each line, and only extension phones could be procured competitively. After a short consideration, the FCC rejected the proposal in July 1978 as inconsistent with the Hush-A-Phone, Carterphone, and equipment registration decisions.[22]

The actual implementation of the registration program for individual telephones was quite straightforward. Each manufacturer produced telephone sets with a standard plug that could be plugged by the consumer into the jack provided by the telephone company, just as an electric appliance is plugged into the electrical outlet. Samples of each model of telephone set were subjected to a series of specified tests to ensure that they met electrical and mechanical specifications adopted by the commission. After passing the tests, the model was assigned a FCC registration number that indicated it was safe to be plugged into telephone jacks.

Incentives to enter the home market for telephones were strong because of the prices charged for extension phones and because of the limited variety in phone styles. Potential entrants recognized the consumer demand for a variety of styles and shapes of phones, as well as for phones with special features. Many companies had the manufacturing capability to produce telephone sets and entry into the home market was rapid. During 1978 phone sets in a variety of styles became available in many department stores and electronics outlets. Approximately one million telephone sets were sold during 1978. The 1979 total was expected to be close to two million.[23] Although the competitive phones were a tiny percentage of the 169 million telephones in the United States in 1979, the rapid sales build-up indicates a likelihood of wholesale replacement of telephone company phones if no deterrent action occurs.

AT&T established two programs to slow the penetration of the competitors. The first was to increase the variety and availability of telephone styles. Novelty and designer telephones were introduced. The company opened phone stores at which a customer could select a telephone, take it home, and plug it in to establish phone service rather than waiting at home for a telephone company repairman to bring a phone and wire it in. A measure of the demand for alternative telephone styles is the fact that AT&T placed 70,000 of its Mickey Mouse telephones while charging $95.00 for the plastic skin alone plus an additional monthly fee for the actual operating parts.[24]

The second tactic was to charge customers for using the telephones they purchased from competitive companies. The original FCC order establishing the registration program provided that customers should notify the telephone company when they plugged registered terminal equipment into the network. The purpose of the notice provision was to assist in diagnosing service malfunctions and to provide statistical information on the extent of interconnected terminals.[25] AT&T then used the notification requirement to impose charges on customers who installed their own telephones. The company continued to impose the previous monthly charge for extension telephones and credited the customer with a lesser amount for providing his own telephone. For example, in Massachusetts the Bell company charged $1.54/month for each residential extension and proposed to credit the customer with $.65/month for providing his own telephone. This arrangement required the customer to pay Bell $.89/month for each telephone purchased from a competitor. The Massachusetts regulatory commission prohibited the charges because the customer-owned extension phones required no new service from Bell.[26] However, most other state regulatory agencies allowed the charges to go into effect. The packaging of the monthly fees on purchased telephones as credits for providing one's own telephone rather than as new charges allowed their implementation without necessarily receiving explicit consideration and approval by state regulatory authorities.

The new charge for purchased telephones would have been an extremely effective method of restricting the new competition if it could have been easily enforced. It allowed the telephone companies to retain charges for extension telephones far above costs so long as the charge was similar to the cost of purchasing a telephone plus the monthly fee to Bell. It avoided the large loss in revenue that would occur from cutting the prices on the 98 percent of the telephones owned by the telephone companies to protect against the new competition. So long as state regulatory agencies accepted the telephone company's charges, the fee could be varied as necessary in future years to compensate for reduced prices of competitors. The ability of one company to charge for all equipment purchased from its competitors is an ideal way to limit the sales of the competitors.

The only flaw in the program was the difficulty of policing the charge. Because the telephone company was not providing any additional service to customers who plugged in purchased telephones, no automatic enforcement mechanism for the charges existed. The existence of extension telephones could be detected by the telephone companies through an electronic analysis of the lines, but such surveillance was expensive and outside the range of routine telephone operations. Consequently most customers who purchased telephones simply plugged them in and ignored the notice that the local telephone company should be notified before using the telephone. In the fall of 1979, AT&T estimated that 80 percent of purchased telephones had not been reported to the local telephone companies. AT&T announced a stepped-up surveillance program in order to detect unreported customer-owned telephones and impose a charge for them.[27]

Implementation of the registration rules for PBX and KTS systems was much more difficult than for ordinary telephones. The registration rules were designed to be fail-safe. Telephones were to contain circuitry to prevent improper voltages or signals from being transmitted into telephone wires regardless of the mistakes made by the installer. For residential telephones, there was little opportunity for improper installation because the only action to be taken was inserting a plug. If the plug was inserted, the telephone was installed correctly; if the plug was not inserted, the telephone was not installed at all. However, PBX and KTS systems were generally installed by skilled craftsmen with extensive onsite wiring. The telephone companies insisted that the fail-safe standards should be applied to KTS and PBX systems, while the new competitors contended that their installers were skilled and competent to perform all the normal wiring functions performed by telephone company installers. Following the breakdown of FCC sponsored negotiations between the parties to determine voluntary standards for the connection of PBX and KTS systems, the FCC instituted formal proceedings in June 1977 to determine the appropriate standards. In April 1978 the commission decided that fail-safe procedures were unnecessary for PBX-KTS systems. The rules adopted required the installer to attest that the work had been done properly and allowed the telephone companies the right to inspect it.[28] Although disputes continued over details of the implementation, the April 1978 order established the basic framework for direct connection of KTS and PBX systems.

The court approval of the FCC registration program in 1977 and the FCC rejection of the primary instrument concept and adoption of PBX-KTS installation standards in 1978 finally broke the tie between terminal equipment and telephone service. Spurred on by advances in technology, strong demand for new telephone styles and features, and attractive telephone company pricing, new competitors expanded rapidly in both the business and residential market after the tie was broken. The low market share of the new competitors as of this writing has made it unprofitable for AT&T to implement the substantial price cuts that would be necessary to stop the competition on a purely economic basis.

If no further legal or regulatory barriers to the new companies are erected, they can be expected to continue increasing their share of the market until the telephone companies find it worthwhile to reduce the price of terminal equipment.

Evaluation of Terminal Equipment Competition

It was twenty years between the 1936 IBM decision that established the general illegality of tying contracts in unregulated industries and the Hush-A-Phone decision that extended the principle to telephone attachments. It was another twenty-two years before the final implementation of the Hush-A-Phone principle in the equipment registration program. The regulatory process maintained a tie between the intrinsically competitive terminal market and the monopoly telephone service market for forty-two years. If state regulatory agencies allow the telephone companies freedom to impose charges on equipment purchased from competitors or to implement below-cost prices for terminal equipment, the tie could be restored.

In the first half of the period, the FCC supported the tying arrangement. The FCC upheld AT&T's refusal to allow the Hush-A-Phone after a long hearing on the matter. The second half of the period consisted of maintenance of the tie through regulatory procedures but without direct support of the FCC. The very restrictive post-Hush-A-Phone tariffs were allowed to go into effect without explicit approval and then found to have been illegal since their inception in the Carterphone decision. The post-Carterphone tariffs requiring protective devices between customer-supplied equipment and the telephone network were allowed to go into effect and later the protective device requirement was found unnecessary and improper. The existence of a regulatory agency provided an impression of public control over AT&T's conduct, but the substance of regulatory control was so limited that the telephone companies were not required to adhere to the regulatory principles. Rather than direct implementation of regulatory rules by agency orders, implementation was left to telephone company determined provisions in the tariffs. This procedure allowed the companies to delay full implementation of the Hush-A-Phone rules for twenty-two years by filing nonconforming tariffs and waiting for the commission to hold hearings, declare the tariff illegal, and order the companies to file another tariff.

From AT&T's point of view, the regulatory strategy was the least expensive method of maintaining a monopoly of the terminal equipment market. So long as a tie could be maintained between telephone service and terminal equipment, entry was blocked into terminal equipment. The tie gave the company freedom to choose the prices of telephone service and terminals in order to take advantage of the interrelated demand without concern for entry in either market. The cost of maintaining the regulatory tie (primarily legal and related expenses for court and commission hearings) was small in relation to the cost of maintaining control of the terminal market through economic means. After the regulatory tie was broken, AT&T found it worthwhile to allow some competition into the market rather than to drop prices far enough to exclude competitors. Further price cuts can be expected in the future if regulatory barriers are not reestablished.

10 New Competition III: Changing Industry Boundaries

Technological Progress and Industry Boundaries

The emergence of competition in long-distance communications and in terminal equipment has been described as a result of a reduction in regulatory barriers to entry. In both cases, the new firm developed a product similar to one offered by the established monopoly firm. The entry was a conventional and expected result of reduced barriers to entry without a fully compensating reduction in the incentives to enter. The new competition examined in this chapter is less conventional. New competition and significant technological progress modified traditional industry boundaries and brought previously unrelated products into competitive interaction.

The three stories of this chapter illustrate three different kinds of changing industry boundaries. The emergence of satellites as a cost-effective method of transmitting communications was a case of a technology used in another industry becoming competitive to existing communications technology. It changed the industry boundaries on the supply side. Aerospace firms with satellite expertise became potential communications entrants. This case fits the second model of Chapter 2. Because the boundary change was on the supply side rather than the demand side, actual development of a communications product and an explicit decision to use the new technology to enter communications was necessary. The satellite case had a similar competitive effect to the development of radio. It did not automatically make the industry more competitive but made large technologically advanced firms potential competitors to AT&T.

In the satellite case, AT&T was dependent on regulatory protection rather than patent protection. When regulatory barriers to entry were removed, the potential competition became actual competition. Satellites extended the effect of radio in breaking down the right-of-way advantages of the established firm. With a single satellite, communications could be established anywhere in the United States. There was no need to develop a vast network of wires or even a network of microwave stations to provide wide coverage.

The competitive interaction of the computer and communications industries provides the second class of industry boundary changes. Technological progress created new products which incorporated a mixture of computer and communications capabilities. It also made computer technology competitive with communications technology and provided potential new entrants as in the satellite case. The final product required computers and communications as inputs. Because firms in both industries expect the new products incorporating combined computer-communications capabilities to be a large portion of the total demand, they have moved toward providing products that combine both inputs. Firms in both industries thus become competitors for the final product, and insofar as they attempt to provide the inputs themselves, they become competitors in the basic products as well.

New technology has reduced the cost of transmitting written messages electronically relative to the cost of transmitting them physically by mail. The change in relative costs is likely to continue and make new forms of telegraph transmission (facsimile and computer-originated messages) competitive with ordinary mail service. If the expected cost trend occurs, the communications companies will come into competition with the U.S. Postal Service (USPS). This is consistent with the third model of Chapter 2. Demand-side industry boundaries change when very large changes in the relative prices of the two products take place. A product that was once confined to a specialized narrow market can become competitive to a broader market if its price declines far enough.

In a free market context, all three types of change discussed in this chapter would lead to greater competition. Regulatory effects make the competitive impact of the technological changes less certain. The competition between mail and communications companies could lead to a legal monopoly by the USPS over portions of the communications market. The convergence of the computer and communications industries could lead to an extension of regulatory control over portions of the computer industry. Evidence so far indicates that the changes will bring greater competition, but the events are just unfolding and the future course of the industries involved is highly dependent on regulatory, court, and legislative decisions of the future.

Satellite Communications

After the successful launching of the Russian *Sputnik* in 1957 and the U.S. *Explorer* in 1958, space technology developed rapidly. The goal of placing a man on the moon before the end of the 1960s combined with the potential military benefits of space technology led to intensive government research efforts on satellites, rockets, and related equipment. Satellites provided a potential new technology for communications. A satellite could operate as an extremely tall microwave tower that could receive, amplify, and retransmit signals to provide communications over a wide area. Although satellite communication was technologically feasible soon after the first satellites were launched, it was far more expensive than existing cable or microwave technology. However, rapidly falling costs of satellite communications could be predicted from the schedule of research and expected accomplishments developed for the moon program. Because space technology was developed by a variety of firms under government sponsorship rather than by AT&T, the declining costs of satellite communications created potential communications entrants.

The potential competitiveness of satellite technology put the communications industry in the situation examined in the second model of Chapter 2. Several noncommunications firms were using satellite technology. The technology was more expensive than the existing communications technology but was dropping in cost more rapidly. Under those conditions, the satellite firms have an incentive to enter when the cost of satellite technology is below the price charged by the monopoly but still above the cost of terrestrial technology. AT&T has an incentive to continue using the old technology until the cost of satellite technology reaches equality with the cost of terrestrial technology, and to continue using terrestrial technology even longer if it must incur significant costs to master the new technology. With equal access to the new technology, the technological progress reduces barriers to entry and prices, but does not necessarily destroy the market-share dominance of the existing firm.

When satellites first became technically feasible, AT&T and the international telegraph carriers proposed that satellite communications be integrated with other forms of communications under carrier control as the technology became economically attractive. In their view, a communications satellite was simply an expensive microwave relay station that was high enough to span an ocean rather than a distance of sixty miles or so under terrestrial microwave technology. AT&T had earlier considered the possibility of using airborne microwave relay equipment to provide two-mile-high relay points but decided against the proposal.[1] The ability to launch a satellite provided a feasible method of obtaining a high relay point that would be particularly useful where standard microwave towers could not be erected (such as across the Atlantic Ocean).

In 1960 AT&T proposed a commercial system of thirty low-orbit satellites that could be operational by 1964.[2] A low-orbit satellite completes an orbit in a few hours and thus appears to move rapidly across the earth's surface. It consequently requires a complex steerable antenna and can only transmit to a given spot for a few minutes in each orbit. At the time of the AT&T proposal, the feasibility of geostationary satellites (which orbit once every twenty-four hours and thus appear to remain stationary over one point) was uncertain.

There was considerable political opposition to AT&T's plan to develop commercial satellite communications as part of its existing monopoly services. Non-AT&T proposals included suggestions that satellite communications be operated by the United States government, by the United Nations, and by a new private company. The ownership question was settled by the creation of the Communications Satellite Corporation (Comsat) in August 1962 by a special act of Congress. Comsat was to be owned 50 percent by the common carriers and 50 percent by the general public with the board of directors chosen jointly by the carriers, the public stockholders, and the president of the United States. Comsat was given a monopoly on American international satellite communications and was restricted to providing bulk communications circuits to other common carriers who would then resell communications services to the public.[3]

Because of uncertainty over the best technology to use, Comsat initially awarded contracts to AT&T and to RCA for the development of low-altitude satellites and also to Hughes Aircraft for the development of a geostationary satellite. The successful launch of the Hughes *Early Bird* in April 1965 proved that the geostationary satellite was feasible and far superior to the low-orbit satellite as a communications device. Commercial transatlantic service was inaugurated with the 240 voice circuits of the *Early Bird.* The success of the Hughes geostationary satellite provided a drastic reduction in the previously expected cost of satellite communications. A single geostationary satellite could provide continuous communications while a large number of low-orbit satellites were necessary in order to always have one in range of the earth station. Earth station costs were greatly reduced by the decreased maneuverability requirements of the large complex antenna. Earth station costs declined from $50 million each for AT&T's 1962 low-orbit *Telstar* satellite to $5 million for geostationary satellite earth stations.[4]

The successful geostationary satellite made international satellite communications competitive with submarine cables at the time of launch. International satellites did not entirely replace international cables because of an explicit regulatory policy that there should be a mix of international communications facilities. New cables continue to be installed even though their cost per circuit is higher than for satellites. Satellite costs were also potentially lower than AT&T's prices for distributing network television programs within the United States. Television distribution made good use of the satellite's technical characteristics, because a single satellite transmission could be received by television stations all around the country.

In September 1965 the ABC television network requested permission from the FCC to launch a satellite that would receive programs from California and New York and transmit the programs to ABC stations throughout the United States. The ABC proposal was a request for a private system rather than for common carrier status. At that time, private microwave systems were authorized but few were built because the Telpak tariff made private systems uneconomical. The application was opposed by Comsat. Rather than ruling directly on the ABC request, the commission instituted a broad inquiry into satellite policies and deferred action on specific proposals until the policy questions were decided. Comsat claimed that the 1962 act conferred monopoly status on it for the space segment of a domestic system as well as an international system and that therefore the commission was without authority to authorize other systems. AT&T and Western Union challenged the economic and technical viability of privately owned systems and advocated a multipurpose common carrier owned domestic satellite system. The television networks supported the commission's authority to authorize private satellite systems and encouraged it to do so.

In January 1970 the Nixon administration sent a memo to the FCC advocating open entry into domestic satellite operations. The memo stated:

> Government policy should encourage and facilitate the development of commercial domestic satellite communications systems to the extent that private enterprise finds them economically and operationally feasible. We find no reason to call for the immediate establishment of a domestic satellite system as a matter of public policy. Government should not seek to promote uneconomic systems or to dictate ownership arrangements; nor should coordinated planning or operation of such facilities be required except as essential to avoid harmful radio interference.
>
> Subject to appropriate conditions to preclude harmful interference and anticompetitive practices, any financially qualified public or private entity, including Government corporations, should be permitted to establish and operate domestic satellite facilities for its own need; join with related entities in common user, cooperative facilities; establish facilities for lease to prospective users, or establish facilities to be used in providing specialized carrier services on a competitive basis.[5]

Two months after receiving the Nixon administration memo, the commission asked for a detailed system proposal from any party interested in participating in the domestic satellite business and deferred the decision on eligibility of entrants until after the proposals were received.[6] AT&T, Comsat, General Telephone and Electronics, Western Union, MCI, Western Tele-Communications, RCA, Hughes Aircraft, Fairchild Industries, and Lockheed all filed proposals for a domestic satellite system either singly or as a joint venture and also asked that other applicants be disqualified. After reviewing the wide variety of applications from communications and aerospace companies, the commission adopted a policy of open entry with some restrictions on existing carriers in June 1972. Three dissenting commissioners approved the idea of open entry but opposed the restrictions placed on Comsat and AT&T. The commission declined to evaluate the demand and limit the approved capacity to the expected demand. It left the responsibility to determine the economic feasibility of the proposals to the various applicants.

The restrictions on the established carriers were designed to prevent them from using their monopoly services to subsidize operations in domestic satellite service. Comsat was required to set up a separate subsidiary to insulate domestic operations from its monopoly international operations. AT&T was allowed to use domestic satellites for its monopoly long-distance services but required to defer satellite service of the private-line market (opened to competition by the Specialized Common Carrier decision of the previous year) for three years after satellite service was instituted.[7]

The domestic satellite decision was a very significant regulatory move toward competition in telecommunications service. The policy of open entry was a sharp departure from the FCC actions in the early days of microwave. Although the private-line market and other specialized services were a small proportion of the total long-distance communications market, the regulatory decision to promote the development of an important new technology outside of AT&T's control provided the potential for much greater competition in later years. The main force behind the commission's changing view of the merits of competition appears to have been the political belief in the virtues of free enterprise and the market system advocated by the Nixon administration. Political influence was exerted through appointments to the commission and direct recommendations by the administration. The political impetus to greater competition was expressed by Dean Burch, Nixon's appointee as chairman of the commission, in the 1973 decision approving GTE's satellite proposal as follows:

> The Commission has been compelled to decide the basic policy issue of GSAT's authorization on fundamental principles. In my case, the overriding principle is that which guides our basic political and economic system: unless there are overwhelming public interest reasons for limiting the freedom of individuals or firms to pursue their own political, social, and economic objectives, based on their own best judgements, such limitations should be rejected . . .
>
> As noted, I have not been convinced by the comparative cost and related economic data presented by either party, nor by their public interest arguments. I am also concerned over the fact that GT&E can expect to realize a return on its investment almost irrespective of its operational effectiveness or economic efficiency—as indeed can Bell.

This could motivate GT&E simply to enjoy the good life provided by its expanded network role.[8]

With the "open sky" decision of the FCC, essentially any company was free to establish a satellite communications network. However, the incentive to do so was limited by the significant economies of scale in satellite communication, the cost of satellite launches, the cost of earth stations, frequency congestion, and the proportion of total telecommunications traffic legally available to satellite competition. Satellite communication economies of scale were much greater than in microwave communication. At the time of the decision, communication satellites contained twelve or twenty-four transponders. Each transponder had communications capacity similar to that of a terrestrial microwave network. To provide adequate back-up facilities, most applicants proposed a system of three satellites, two in orbit and one as an on-ground spare. Entry on a small scale was infeasible.

The cost of satellite launches and of earth stations was interrelated. The satellite needed a large enough antenna and sufficient power to receive the signal from the earth station, amplify it, and retransmit it with enough power to be received by another earth station. A satellite with a large antenna and extensive solar cells to provide high power could operate with relatively inexpensive earth stations. A smaller satellite with less power required much larger and more expensive earth stations. A tradeoff also existed between the capacity utilization of the satellite and the size of the earth station. Reducing the communications volume on a satellite of a given size allowed a reduction in the size of earth stations.[9] Receive-only earth stations were much cheaper than send-receive stations, reducing the earth station costs for television network distribution (requiring one transmit and many receive-only stations) relative to point-to-point communications (requiring send and receive at each point).

The potential tradeoffs between satellite size, capacity, and earth station costs made cost minimization a complex decision dependent on the particular characteristics of the planned communication network. With low launch costs, a large high-powered satellite could provide communication among many inexpensive earth stations and give much greater flexibility in reconfiguring communications paths than existed with cables or microwave. The high launch costs at the time of the domestic authorization restricted the economic feasibility of satellites to low-power satellites with large expensive earth stations. Thus satellites could be used to provide service between distant cities with dense two-point communications (such as New York–Los Angeles) but required supplementary terrestrial facilities to provide complete coverage.

An additional limitation on satellite communications was frequency congestion. Initial satellite frequency allocations were in the same bands as terrestrial microwave, which were quite congested near the largest cities. Earth stations consequently had to be placed some distance outside the city and terrestrial facilities used to relay communication into the city. Satellite carriers depended on established telephone companies for local distribution, subjecting them to competitive use of the connecting rights, and limiting them to the technical capabilities of established facilities. Much greater use could be made of satellites if direct distribution to large users was possible. The use of higher frequencies (10 GHz or more instead of the 4- to 6-GHz initial bands) could relieve the frequency congestion and also allow smaller earth stations (all other things equal, the higher the frequency, the smaller the necessary antenna to receive it) with a consequent reduction in costs. However, the use of higher frequencies required technological advances from the equipment then in common use. Higher frequencies were also more affected by weather conditions than lower ones.

The final limitation on satellite effectiveness was the size of the market legally open to competition. Prior to the Execunet litigation, the switched long-distance market was considered a legal monopoly of AT&T. Consequently the new entrants were limited to television distribution, private-line service, and any new services they could develop. The legal limitations increased the significance of economies of scale and high earth station costs as barriers to satellite development. Earth station costs restricted satellite service to a small number of large cities, while legal limitations prohibited new entrants from carrying switched long-distance traffic between those points and economies of scale required that they carry large volumes of communications in order to use capacity efficiently.

Even with the ability to use its satellites for switched long-distance services, AT&T did not consider satellites to be cost effective with terrestrial systems at the time of the domestic authorization. The company estimated that the annual cost of operating a three-satellite system (including earth stations) would be $65 million/year while the annual cost of equivalent terrestrial facilities would be $60 million/year. However,

AT&T still made plans to develop a satellite system in order to provide additional information and network flexibility.[10] AT&T joined with Comsat General (the domestic satellite subsidiary of Comsat) and General Telephone to develop a single system owned by Comsat and leased to AT&T and General Telephone. The consortium launched two satellites in 1976.

Four companies established competitive satellite service on two different satellite systems. Domestic satellite service was inaugurated in January 1974 by a subsidiary of RCA. RCA had previous communications experience as an operator of international telegraph cables and the communications system of Alaska, as well as satellite technology expertise. RCA initiated service by leasing channels in bulk on a Canadian satellite. The company launched its own satellite (built by RCA) in December 1975 and followed with a second satellite four months later. Western Union's satellite subsidiary launched its first satellite in April 1974 and its second soon afterward. American Satellite Corporation, a subsidiary of Fairchild Industries, abandoned its original plan to launch a very large satellite based on the Applications Technology Satellites that Fairchild built for NASA. Instead American Satellite began service in August 1974 by leasing satellite capacity on the Western Union *Westar* and building its own earth stations. Southern Pacific Communications also entered the market by leasing satellite capacity and establishing its own ground facilities. Southern Pacific's entry into satellite communications and its purchase of the bankrupt Datran network of digital data transmission facilities, along with its existing terrestrial microwave network, gave the company wide flexibility in configuring communications services to meet the demands of its customers.

The competitive segment of the satellite communications market during the AT&T moratorium period consisted of two satellite systems with four final sellers. Three basic services were offered: (1) television transmission service from one point to receive-only stations located near individual television stations, (2) private-line voice circuits tied into the AT&T network, and (3) specialized services customized for individual companies. By total capacity utilized television transmission was the most important service offered. The new companies were particularly successful in interconnecting cable television networks. The cost of receive-only antennas dropped rapidly as they came into widespread use and as small antennas were found to provide adequate reception of television programs. Applications for television receive-only earth stations prior to actual satellite operation proposed antennas of about 35 feet in diameter at costs ranging from $100,000 to $150,000 per station depending on the location and the applicant. By the end of 1979, receive-only earth stations were being built with antennas fifteen feet or less in diameter at a cost as low as $10,000 each. By October 1979, 2,250 receive-only earth stations had been built for cable TV networks. At that time the mandatory licensing requirement of receive-only stations was removed allowing users the option of building stations without the delay and expense of an FCC license if they were willing to forgo protection from potential radio interference. At the end of 1978, eighteen of the twenty-four transponders on the first RCA satellite were devoted to cable television transmission service.[11]

The second class of service was private-line service similar to that offered by the terrestrial microwave carriers. All four retail carriers offered practically identical terms based on the prices charged by terrestrial carriers rather than on satellite costs. Each of the carriers built large earth stations at a few selected points across the country (such as Los Angeles, New York, Chicago, and Houston) and provided service to other locations through smaller earth stations and/or microwave links. By 1978 all four carriers had settled on a basic channel charge of $1,000/line/month for transcontinental routes (New York–Los Angeles) and $500/line/month for shorter routes served by the carrier (such as Chicago–Houston). Prices varied between the $500 minimum and the $1,000 maximum for routes between one thousand miles and transcontinental distance. As a comparison with terrestrial carriers, the channel charge per line per month for MCI was $1,270 for Los Angeles–New York and $664 for Houston–Chicago. The corresponding charges under AT&T's MPL tariff were $1,346 for Los Angeles–New York and $726 for Houston–Chicago.[12] Thus the satellite carriers' price was less than the MCI or AT&T price for both routes. The satellite carriers did not have exactly the same rates to final customers because they varied in miscellaneous charges such as channel termination, local distribution, and installation charges, as well as in the discount structure for groups of lines.

Because the cost to the carriers is the same for transmitting from Chicago to Houston as from New York to Los Angeles, the price structure was notable for its maintenance of non–cost-based rates even with four carriers. We ordinarily expect non–cost-based rates to be competed away in the absence of collusion or monopoly power. The price structure can be explained as a function of the loading of the satellites. So long as the satellites were below capacity loading, the marginal cost to the carriers of adding an additional circuit was close to zero. The costs of satellites, launching service, and earth stations were all sunk costs and any action that would increase total revenue would be profitable. Thus a firm would find it worthwhile to offer short-distance satellite circuits at a price below the average cost in order to increase the satellite fill. Reduced prices for short routes could also increase the demand for profitable long-distance circuits. Because many companies with facilities on both coasts have plants in the Midwest as well, they are more likely to accept a satellite communication system if service is offered to all of their plants than if it is offered to only the most separated ones.

The third class of services offered by the satellite companies was specialized services for individual customers. American Satellite was particularly active in pioneering private networks using small earth stations at or near the customer's premises. The first important private network was set up by American for Dow Jones to transmit the *Wall Street Journal* to regional printing plants. The system included three private earth stations.[13] Other private networks were designed for Sperry Univac, United Press International, Associated Press, and Muzak. The private networks were not an important source of revenue during the AT&T moratorium period, but they were significant for providing experiments with innovative services and small earth stations.

By 1979 the first phase of domestic satellite service was completed. All four satellites of RCA and Western Union were filled close to capacity. Considerable unsatisfied demand from cable television companies existed. Western Union launched its third satellite in August 1979. RCA launched its third satellite in December 1979 but lost contact with it soon after the launch. The $50 million investment in the RCA satellite and launch costs was lost. Because of long lead times for producing satellites and arranging launches, RCA did not expect to replace the lost satellite until 1981.[14]

The AT&T–GTE–Comsat consortium launched its third satellite in June 1978. The launch of the third AT&T satellite generated considerable controversy because of the underutilization of the first two and the scheduled expiration of the moratorium on use of the satellite for competitive purposes. Launch authority was requested just prior to the launch date after all launch arrangements had been made. The FCC Common Carrier Bureau recommended against allowing the launch because the first two satellites were only being used at 10 to 15 percent of capacity. The commission approved the launch because denial would cost $8 million to remove the spacecraft from its launching pad and reschedule a launch at a later date.[15] The AT&T launching was of concern to the satellite competitors because of the expiration of the moratorium on AT&T provision of competitive satellite services in July 1979. If AT&T had substantial excess capacity with practically zero marginal costs, it could make a strong case for very low priced competitive satellite services.

The first phase of domestic satellites clearly demonstrated the economic feasibility of using television distribution via satellite, but had little impact on the point-to-point voice and data market. The expected completion of the space shuttle to reduce launch costs and improvements in technology to utilize higher frequencies suggested the possibility of a much greater satellite impact in the 1980s with a large high-powered satellite and small inexpensive earth stations near users' premises. The existence of the free-entry policy and satellite expertise outside of AT&T made it likely that extensive satellite competition would take place. The expiration of the AT&T moratorium and AT&T's possession of substantial excess capacity also promised a replay of the pricing controversies of the terrestrial microwave market. However, AT&T's freedom to make narrowly focused competitive price adjustments was more limited in the satellite case because of the greater ease of reconfiguring a satellite network and the Execunet litigation that broadened the scope of competition to the switched long-distance market.

Computers and Communications

The fundamental components of a communications system are line haul equipment, switching systems, multiplexing systems, and terminal equipment. Satellite transmission affected only the line haul component. However, the other three components were threatened by technological advance in the computer industry. The competition of computers with communications fits the same theoretical model as satellite competition. Technology developed for a different industry enjoyed rapid technological progress and consequently became competitive with traditional communications industry technology. The technological progress brought computer firms into the communications industry and caused AT&T to adopt the new technology for its communications operations and to expand its operations into the previous domain of computer processing.

A basic computer consists of a central processing unit (CPU) connected to a variety of peripheral devices (tapes, disk drives, card readers, printers). The system is functionally equivalent to a communications system. Consider, for example, the simple listing of punched cards on a printer. Information on the cards is transformed into electrical impulses by the card reader (as the voice is transformed into electrical signals by the telephone set), the electrical signals travel over a connecting cable to the CPU (as the telephone signals travel over a wire to the telephone office) where the signals are formatted and switched to the cable connecting the printer. Because one channel often has a much higher data rate than terminal devices, signals are often multiplexed on the channel just as voice signals are multiplexed on a long-distance circuit. Data may be switched and multiplexed more than once, just as a telephone message may be processed through more than one central office. The transfer of data from a disk drive to a tape drive generally requires movement from the disk drive to the disk controller where it is multiplexed with other data on the channel connecting the controller and the CPU (equivalent to a telephone interoffice trunk). The data then travels to the tape controller (equivalent to the destination central office) where it is switched to the destination tape drive.

Because computer systems incorporate communications functions, computer technology can conceivably displace traditional communications technology for switching and multiplexing, and to a lesser extent for terminal equipment. However, major differences in the design goals of the computer and communications industries slow the displacement process. The most significant difference is that computers are designed to transmit data and telephone systems are designed to transmit voice. Data takes on discrete values and can be coded precisely in digital form. The switching and multiplexing activities in computers are consequently designed to process digital signals. Voice takes on infinite gradations. The telephone system consequently handles signals in analog form (varying continuously) and its equipment is designed to manage analog signals. Significantly different switching and multiplexing technology is used to control analog and digital signals.

A second difference is in the requirements for accurate transmission. Data transmission requires extreme accuracy. The improper transmission of a single bit can lead to a significant change in the results. Digital transmission is suited to high accuracy because the receiving station need only distinguish between two states for each bit in order to regenerate it exactly. In addition, the discrete nature of digital signals allows the use of check bits that bear a particular mathematical relationship to groups of bits and indicate an error condition when a bit is received improperly. Voice transmission is satisfactory with some errors. Transmission errors may blot out whole syllables without destroying the intelligibility of the speech because of the ability of the listener to reconstruct missing parts from the context. Analog signals provide less accuracy than digital signals because they may be amplified but not regenerated.

A third difference between the design goals of the two technologies relates to the problems encountered from entire system failure. In early stand-alone computers, an entire system failure simply meant the loss of computer capability for a time. While it was obviously not welcomed, a system failure was often less serious than the undetected transmission of faulty data. Because of the public nature of a telephone network, an entire system failure eliminates service to many people. Even serious errors in transmission of individual conversations are preferable to total system failure. This difference has become less significant in recent years because many computers support large time-sharing networks and take on a public nature similar to the telephone network.

The differences in design goals between the two technologies allowed each to exist independently under a wide range of relative prices. In the early years of the computer industry, it was economically efficient for data to be handled in digital form on computers with accurate individual transmissions but subject to system failure, while voice was handled in analog form and switched on highly reliable electromechanical switches. However, with sufficient cost differentials, either technology could displace the other. Voice can be converted into digital form by sampling the voice signal very frequently. A very high bit rate is necessary to transmit subtle differences in voice accurately, but with sufficiently low-cost computer technology, digital transmission of voice is cost effective. The differences in bit rate necessary to transmit the information contained in conversation and the actual voice are vast. The actual words transmitted at normal conversational speed could be carried at about 200 bits per second if the words were coded in standard computer form. Accurate digital transmission of the voice under current methods requires 56,000 bits per second, because the voice is sampled 8,000 times per second and each sample is encoded in seven bits.[16]

Digital computer technology has dropped in cost more rapidly than analog voice technology and led to incorporation of digital components into the voice network. The substitution of computer technology began with the use of a special-purpose computer developed by AT&T for the control section of a large central office switching system introduced in 1964. That system continued to use electromechanical switches for the actual circuit switching function. Simultaneously the development of time-sharing computers and increased use of computers to access centralized data bases led to geographical separation of the final user and the CPU. Public telephone lines were then used to connect the terminals and the CPUs. While the telephone lines were functionally identical to the cables connecting the CPU and peripheral devices in stand-alone computers, the use of public lines expanded the computer system to include parts of the telephone system as well.

The increased use of computer technology in the telephone system and increased use of telephone lines in computer systems brought the two industries into close interaction. Several possibilities existed for the relationship between firms in the two industries: (1) they could engage in direct competition with computer companies using their technological capabilities to enter the communications industry and the communications carriers expanding their activities to include data communications networks as well as voice networks; (2) either industry could supply only components to the other and develop their traditional technologies; and (3) new firms could be formed that would purchase inputs from both industries and offer computer communications products to the public. The situation was complicated by the existence of regulation for the communications companies but not for the computer companies and the restriction of AT&T to regulated activities through the 1956 Consent Decree. If regulatory control could be extended to data processing services, AT&T would have an opportunity to control that market through its control of the communications lines, prohibitions on resale of those lines, and familiarity with the regulatory process. If the regulatory line could be drawn narrowly around traditional communications, AT&T could be prohibited from entering the combined services by the Consent Decree.

In the early 1960s, Bunker Ramo provided a stock quotation service to brokerage offices using its own computers and leased AT&T lines. The service was treated as an unregulated data processing service. When Bunker Ramo added a message switching capability to its service in 1965 to allow brokers who used the stock quotation system to transmit buy and sell orders, AT&T refused to lease the necessary communications lines. According to AT&T, the addition of message switching made the service communications rather than data processing and thus constituted unallowable resale of AT&T-provided lines. The resulting Bunker Ramo complaint caused the FCC to initiate an inquiry into the distinctions between data processing and communications service and into what part of the services ought to be regulated.

After five years of consideration, the FCC promulgated its computer rules in 1971. Services were divided into four categories: data processing, hybrid data processing, hybrid communications, and communications. The regulatory line was drawn between hybrid data processing and hybrid communications. Communications and hybrid communications were regulated and hybrid data processing and data processing were unregulated. The distinction between hybrid data processing and hybrid communications was to be drawn by relative use; if the primary use of the service was communications it was regulated hybrid communications. If the primary use was data processing the service was unregulated. Regulated common carriers were prohibited from providing data processing service except through separate subsidiaries. AT&T could not provide data processing service at all under the decision because it was restricted to regulated communications services by the 1956 Consent Decree.[17]

The 1971 rules decided little. They left an arbitrary line between data processing and communications that was not based on underlying technological and economic forces. Services that would naturally have been offered as a unified service in a free market were either not offered at all or were split into two distinct services, one part regulated as a communications service and the other unregulated as a data processing service. Continued problems in applying the 1971 rules to actual and proposed services caused the FCC to undertake another inquiry into the proper dividing line in 1976.

Near the beginning of the second computer inquiry, the FCC completed an investigation of the resale restrictions of the telephone carriers. The resale restrictions were a fundamental plank in the carriers ability to impose discriminatory pricing schemes because otherwise favored customers would resell to less favored ones. Although restrictive provisions had been standard practice since the creation of the FCC, they had never been formally evaluated and approved. The proceeding grew out of a desire by certain companies to lease lines from AT&T and to "add value" to those lines by computer capabilities to provide a new value-added communications service. The commission ruled that the resale and sharing restrictions were unlawful discrimination and should be removed for all services except MTS and WATS. The fundamental legal principle underlying the decision was a 1911 Supreme Court decision which prohibited the railroads from refusing service to freight forwarders who purchased railroad service in bulk (carload lots or greater) and resold it to smaller shippers.[18] Because that decision interpreted the Interstate Commerce Commission Act on which the common carrier portion of the Federal Communications Commission Act was based, the FCC ruled that the reselling of communications service was analogous to freight forwarding and could not be prohibited by the carriers.

In the resale decision the commission made a distinction between resale and sharing. Resale carriers were subject to regulation, but with a policy of fairly free entry. Entrants would have to show they were legally, technically, and financially qualified to offer the proposed service, but the services would be presumed to be in the public interest and no issues of economic harm to the existing carriers would be considered. Rate regulation and exit regulation would be imposed. Sharing arrangements would not come under regulation, whether performed in a pure sharing mode or via a nonprofit intermediary. The commission went to some length to distinguish sharing arrangements from resale arrangements, but the rules left considerable room for overlap of the two categories. AT&T vigorously opposed the rulings and gained several delays of their implementation but was unsuccessful in its attempt to overturn the resale rules. The appeals court ruled in favor of the FCC and the Supreme Court refused to review the case in 1978, completing the proceeding.[19]

The resale decision to allow competitive, regulated carriers who resold communications service was used as the foundation of the tentative decision in the computer inquiry. In the July 1979 tentative decision, the FCC established four categories of services: basic voice, basic nonvoice, enhanced nonvoice, and data processing. Basic voice and basic nonvoice services were the fundamental communications services in which the form of the message is the same when delivered to the carrier and when received by the final customer. Basic voice and basic nonvoice services were to be provided by regulated common carriers in the traditional manner. Enhanced nonvoice services consisted of all services that were largely communications but that included data processing as well. Enhanced nonvoice services were to be separated from basic services under the tentative decision, but to remain regulated. Enhanced nonvoice services could be provided either by carriers who leased basic facilities or by underlying carriers who owned basic facilities. If provided by underlying carriers, the enhanced services were to be provided by a separate subsidiary from the basic services. Data processing service could be provided on an unregulated basis by the same entities that provided enhanced nonvoice services under regulation, whether they were separate companies or subsidiaries of underlying carriers. Ordinary telephones and other traditional communications terminals could continue to be offered on a tariffed regulated basis. Terminals with additional capability could either be offered on an unregulated basis or on a regulated basis via an enhanced nonvoice company or subsidiary, but not by an underlying carrier.[20]

The tentative decision expanded the potential scope of regulation into many services that had previously been considered pure data processing. For example, a standard procedure of time-sharing computer service companies was to lease private-line circuits from the carriers to establish a data communications network to remote locations. Customers could then access the central computer via a local call to the local office and over the data processing company's leased line to the central computer. This was considered data processing service and not resale of communications service. However, the new rules left open the possibility that services of that type could come under regulation if the customer primarily retrieved messages from the central computer rather than performing extensive processing.

The vagueness, complexity, and scope for regulatory expansion contained in the proposed rules provoked widespread dissatisfaction. Consequently the final decision (April 1980) was a substantial departure from the tentative decision. The final decision was a clear move toward deregulation. All terminal equipment (including basic telephones) was to be deregulated as of March 1982. After that time, telephone sets could only be provided by AT&T or GTE through a separate unregulated subsidiary, not as part of the basic telephone service. Only basic transmission would remain regulated. Transmission services enhanced with computer processing could only be sold by AT&T or GTE through separate subsidiaries that purchased basic transmission capacity on a tariff basis. Smaller telephone companies were allowed greater freedom in structuring enhanced services and terminal equipment offers than were AT&T and GTE.[21]

The final decision will be an extremely significant event in the telecommunications industry if it is implemented. However, implementation will not be easy. The decision cut across many established regulatory patterns including the treatment of terminal equipment in the separations formulas and the restriction of AT&T to regulated services by the 1956 Consent Decree. The FCC declared that AT&T would not be barred from the newly deregulated areas (terminal equipment and all enhanced communications services), but only the Justice Department and the court can make a definitive interpretation of the decree. Thus it is possible that the decision would bar AT&T from large areas of communications services if a strict interpretation of the Consent Decree is followed.

State regulatory commissions and small telephone companies are likely to strenuously oppose the deregulation of terminal equipment and the resulting changes in the payments for long-distance calls between AT&T and the small telephone companies. Some new entrants or potential entrants to portions of the communications industry who had advocated deregulation were concerned that inadequate provision had been made to prevent AT&T from subsidizing unregulated competitive services with revenues from regulated monopoly services. The widespread scope of the decision and its failure to protect the status quo led to court challenges immediately after it was rendered. *Telecommunications Reports* described the decision as potentially "the biggest legal battle of this decade."[22] The expected long legal battle and the property rights in the status quo inherent in the regulatory process make it unlikely that the major changes contemplated by the second computer inquiry final decision will be implemented on the announced schedule. Thus even after fourteen years of consideration, the regulatory treatment of the boundary between computer and communications services was uncertain in 1980.

The computer inquiry prior to the disputed 1980 final decision left the regulatory boundary vague and subject to individual interpretation. Consequently, companies from both industries interpreted the line to their benefit. AT&T offered products combining communications and data processing as regulated communications services. Data processing companies offered similar services as unregulated data processing services. For example, PBXs are traditional parts of the regulated communications service. When AT&T introduced its Dimension PBX line, which was based on digital computer technology, it also gained a capability to perform standard data processing functions that was absent in the previous electromechanical PBX systems. AT&T consequently introduced a hotel/motel feature on the Dimension that allowed it to perform certain accounting functions previously handled by separate computers. Similarly AT&T had long supplied teletypewriter terminals under tariff as part of its communications services. When considerable computer "intelligence" was built into an upgrade of the traditional teletypewriter line, it became competitive with intelligent terminals offered by unregulated computer companies. AT&T also announced a new enhanced communications system for computer communications that would allow communication between incompatible terminals and perform functions ordinarily done by private computers on other networks, but programming difficulties delayed the actual introduction of the system. These and other new services combining computer and communications capability were protested by data processing firms as falling outside the regulated communications market, but without any clear results.

While AT&T moved into traditional computer functions, IBM and Xerox made plans to offer direct communications competition based on their computer expertise. The combination of satellite technology for line haul and computer technology for switching, multiplexing, and terminal equipment offered the possibility of setting up a complete communications system separate from the traditional analog communications technology.

To develop new kinds of integrated data and voice service using digital technology, high-powered satellites, high frequencies, and low-cost earth stations at customer premises, IBM proposed a joint IBM–Comsat General satellite subsidiary in 1974. The proposal met great opposition from both communications and computer companies. It was attacked as anticompetitive in the computer industry, because the new system could be tailored to the requirements of IBM computer systems and be used to gain an advantage over other computer manufacturers. It was challenged as anticompetitive in the communications industry because it was a joint venture of two companies that could potentially enter the satellite communications market separately. The commission favored the general proposal but sought to allay fears of competitors by imposing a number of conditions on the venture. The partnership was required to publish standards and provide interconnection with other systems on a nondiscriminatory basis. The IBM parent corporation was prohibited from promoting directly the services of the communications subsidiary. The commission also required a restructuring of the venture to reduce IBM's control. The partnership added Aetna Casualty and Surety Company and named the consortium Satellite Business Systems (SBS). SBS filed formal applications for a satellite network in December 1975 and after additional commission consideration, the plan was approved in January 1977.[23]

Opponents of the SBS plan, including AT&T, Western Union, American Satellite, and the Department of Justice, then appealed the FCC decision to the appeals court. In August 1978 the appeals court reversed the FCC and remanded the case to the FCC for evidentiary hearings. The court found that the FCC had not adequately examined a potential Clayton Act violation of the joint venture in which it was possible that one or both of the parties would enter the market separately. Rather than beginning the evidentiary hearing ordered by the court, the FCC and SBS appealed the decision to the full appeals court. In May 1979 the full court vacated the previous order remanding the case to the FCC and scheduled oral argument. The September 1979 hearing focused on procedural rather than substantive issues, examining the FCC's ability to make various kinds of competitive judgments without evidentiary hearings. The FCC contended that a hearing would take "several years at a minimum" and that the applications should not be delayed that long.[24]

Although final approval of the system had not been obtained as of this writing, it is generally expected that the system will be built, either in its currently proposed corporate form or with some restructuring imposed by court orders. During the litigation, SBS has continued developmental work. Because the original plans called for extensive development of new equipment and delay of operational status until the early 1980s, the regulatory delays have not held back the program to a significant degree. The FCC continued to authorize SBS preparations as if the system were approved. In April 1979 the FCC approved an SBS plan to build two tracking, telemetry, and command stations at a cost of $4 million each. SBS continued developmental work on the system, including contracts with other companies for equipment to be used with the SBS satellite system. In February 1979 SBS selected AM International (formerly Addressograph-Multigraph) to develop a high-speed copier-facsimile machine that would have the capability to transmit 3,600 pages/hour via the SBS satellite links, 120 times the rate of common 30-page/hour facsimile machines designed to transmit over ordinary telephone lines.[25]

The SBS program was designed to provide a single communications system for the transmission of voice, data, facsimile, and video signals. All transmission would take place in digital mode. Multiplexing would be by time division rather than by space division. The system was designed to operate in a similar manner to a large time-sharing computer that allocates time in tiny slices to many different terminals, giving each user the impression of instant response. Each earth station was to be polled fifty times a second and allocated a few milliseconds each time for the transmission of signals. The signals would then travel together to the satellite and back again where they would be separated in sequence at the destination locations. The use of digital transmission and time-division multiplexing allowed the direct application of computer technology and techniques to a communications system.

SBS proposed to use frequencies in the 12-GHz range and to use small earth stations at the individual customers' premises. The earth stations were expected to cost about $250,000 each, including controller electronics. Thus the initial proposal would only be useful for users with a very high volume of communications. SBS estimated that 415 companies were potential customers and worked out design plans with sixteen companies as part of its preparation for the system.[26] It is likely that considerable time will be required before SBS becomes a major force in communications because of the limited number of customers for which the private earth stations would be feasible at the originally planned costs. However, it is likely that the costs of earth stations will decline substantially with improvements in computer technology. Thus the long-range implications of the SBS proposal are for a strong competitor for long-distance traffic that is not subject to defensive AT&T moves on local tariffs. An additional implication of the proposal is that the earth stations could be set up quickly in new locations to provide competitive long-distance service if the price structure of voice communications made it profitable to do so.

A second proposal with significant potential to increase competition via satellite communications came from Xerox at the end of 1978. The proposed capabilities of the Xerox network were similar to those of the SBS network. It was based on unified digital transmission of voice, data, and facsimile to completely bypass the existing interstate and local telephone network. Rather than proposing satellite receivers for each customer, Xerox proposed a radio-based local distribution network with transmission in the 10-GHz range. Xerox would transform all communications to digital mode at the user location and beam the signals to a metropolitan earth station via the 10-GHz microwave link. Consequently customers with too low volume to economically use a private earth station would gain access to the system while avoiding the local distribution problem of the existing satellite carriers. Xerox proposed leasing satellite capacity from other carriers rather than launching its own satellites. The estimated investment required for the proposed system was $500 million.[27]

Mail and Communications

Telegrams and mail service have always been somewhat competitive. The earliest telegraph market consisted of displacing the mail service for users who placed a high time value on their written communications. But the large price differential between the two services has kept them in separate markets. Because telegraph prices have remained a substantial multiple of mail prices, moderate price changes in either market have little impact on the other. A drastic reduction in the price of telegraph service would allow it to take over a substantial portion of the mail market as examined in the third model of Chapter 2. It would not totally displace mail even if its price were lower because physical transmission has characteristics that are more valuable to some customers than delivery speed.

Traditional telegraph service has remained high in cost relative to mail despite extensive technological progress in telecommunications because of the high labor content involved. Messages must be typed in a teletypewriter and then received, sorted, and delivered at the other end. Even if transmission costs declined to zero, substantial total costs would remain. However, advances in facsimile technology and the development of computer-originated mail promise a significant reduction in the cost of transmitting written messages electronically. If the expected cost savings materialize, electronic mail could become competitive with physical mail. Such a development could put the communications companies into direct competition with the U.S. Postal Service (USPS). The expectation that an extensive market for electronic mail will develop has already led to extensive legal and regulatory maneuvering for advantageous position in that market.

With facsimile transmission, a facsimile machine on one end scans the input document and translates it into a series of electronic signals. The scanner does not recognize and transmit individual characters but patterns of light and dark spots on the paper. Thus it can transmit drawings or figures as well as written text, but requires a large volume of signals to transmit a single page. The signals are transmitted over ordinary telephone or telegraph lines and translated back into the image of the input document at the other end. Facsimile has been used since the 1920s for transmitting weather maps and news photographs. However, such machines were expensive and required long transmit times with resulting high line charges. They were consequently uneconomical for transmitting routine business documents.

The business facsimile market can be dated from the introduction of the Xerox Telecopier I in 1966, which was followed by the Telecopier II (1968) and III (1970). The Xerox line was designed to provide economical and convenient transmission rather than the highest possible quality. The Telecopier III could transmit an $8\frac{1}{2}$-by-11-inch document over ordinary telephone lines in four to six minutes. Four-minute transmission was accomplished with a horizontal scan of 96 lines per inch and a vertical scan of 64 lines per inch, while six-minute transmission increased the vertical scan to 96 lines per inch. The greater the scan rate, the better the resolution of small letters and figures, and the greater the amount of data to be transmitted per page. Because good resolution of characters requires ten to twelve scans per character, the Xerox unit could provide good transmission of characters as small as one-eighth inch, but would not provide the fine detail needed for news photos and weather maps. The Telecopier III could be purchased for just under $1,000.[28]

During the 1970s a wide variety of business facsimile machines were introduced. Xerox upgraded its original line with the TC 400 in 1970, the TC 410 in 1973, and the TC 200 in 1975. The TC 200 employed a laser scanner for facsimile input and a Xerographic process for output. It could transmit as fast as two minutes per document. Graphic Sciences, a company formed by former Xerox people to produce facsimile machines in 1967 and acquired by Burroughs Corporation in 1975, introduced a wide line of business facsimile machines. The 3M Corporation also became a major competitor in facsimile after acquiring the Magnafax Division of Magnavox and establishing marketing rights for machines manufactured by Matsushita Graphic Communications of Japan. Qwip Systems, a subsidiary of Exxon, entered the market in 1974 with a $29/month unit that was compatible with the Xerox four- to six-minute machines. The low price generated fast market acceptance of Qwip and made the company the fastest growing facsimile competitor. Several additional companies introduced business facsimile models during the 1970s.

By 1977 the business facsimile market had become much larger than the original news and weather facsimile market. Business documents accounted for about 80 percent of the total facsimile market. A total of 120,000 business facsimile machines were in use. The 1977 sales were $110 million. Xerox was the leading firm with a little less than 50 percent of the installed base, followed by Graphic Sciences and 3M, each with 15 to 20 percent and Qwip with about 13 percent of the installed units.[29]

With facsimile, tradeoffs exist between machine cost, communications line cost, and quality of transmission. The basic facsimile process is intrinsically a very inefficient user of communications capacity. An ordinary business letter consists primarily of blank space (margins and space between lines). A device with a resolution of 96 lines per inch vertical and horizontal will transmit 9,216 samples per square inch, whether the square inch is fine detail or blank space. With increased facsimile machine cost, the process can be compressed so that the information actually transmitted is real information and not simply repeated indications of blank space. Very high quality transmission requires fine spacings of the scan, while adequate reception for many purposes can be achieved with less frequent spacings.

A systems effect exists in facsimile machines similar to that in telephone systems. A single facsimile machine is of no use. The value of having a facsimile machine depends on the number of other people or businesses who have the machines. Even if the costs of machines and communications lines are low enough to promote widespread use of facsimile in equilibrium, reaching the equilibrium may require considerable time because initially there are few opportunities to communicate. To solve the problem of demand for facsimile before the widespread availability of private machines, Xerox established a group of facsimile service bureaus (Teleservice) to serve customers without their own facsimile machines. Graphnet began service in 1975 as a "value added" common carrier to transmit facsimile messages for a fee. The Graphnet price schedule was below the comparable cost of public message telegrams but well above the cost of mail service.[30]

The substantial improvements in voice-line–oriented facsimile machines, the growth of service bureaus such as Xerox's Teleservice and Graphnet's service, and the predictions of high-speed, low-cost facsimile transmission from new services such as the SBS 60 pages per minute satellite transmission have created the possibility of a significant diversion of traditional mail service into facsimile service. The potential competition between electronic mail substitutes and ordinary mail was not missed by the Post Office. During the 1970s the Post Office undertook both in house and outside contractor research on various means of establishing electronic mail service. These considerations involved both facsimile schemes and computer-originated mail which would be printed out at post offices. The latter offered the most promise of economical implementation because it was cheaper to transmit the limited data for printouts than the data for facsimile and thus more competitive with other mail services.

As the first step in its entry into electronic mail, the Post Office signed a contract with Western Union in 1978 for joint provision of Electronic Computer Originated Mail (ECOM). ECOM was similar to the earlier Western Union Mailgram service in which Western Union transmitted messages electronically and the Post Office made final delivery through the mail. ECOM required all messages (two-page maximum) to be submitted by computer to Western Union facilities in batches of at least 200. The messages would then be transmitted over Western Union's regular network of lines and switches to twenty-five receiving Post Offices where they would be printed, inserted into envelopes, and added to the first class mail stream for ordinary mail deliveries. The price for using the service was from $.30 to $.55 per message depending on the total volume of messages sent by a customer. The promise that messages would be delivered within two days gave the service better speed than ordinary mail, but hardly competitive with other types of electronic communications.[31] The restriction of input to a minimum of 200 messages per batch along with other volume restrictions and volume discounts restricted the service to large mailers. The characteristics of the service thus marked a significant departure from previous Post Office flat-rate price structures.

Although the technology and function of the service was similar to

Mailgram, the regulatory and market structure implications of ECOM were far different. Mailgram was a continuation of Western Union common carrier telegram service, but with delivery through the mail providing lower cost and slower transmission speed. The proposed ECOM service was to be a subclass of first class mail under the complete control of the Post Office. The Post Office was to take responsibility for the price and marketing of the service and Western Union to be an input supplier. The service thus raised the question of whether the FCC or the Postal Rate commission (PRC) would regulate electronic mail and whether the private express statutes that guaranteed a Post Office monopoly of letters would be applied to electronic mail as well.

In September 1978 the USPS submitted the proposed ECOM service to the PRC for approval. Western Union did not file an FCC tariff for its part of the service, asserting that it had made a private contract with a single customer and was not offering common carrier service to the public. The Post Office planned to begin the service December 15, three months after the announcement. Selected customers had already indicated plans to use the system prior to the public announcement and request for PRC approval.

The ECOM announcement created considerable controversy and a very unsettled regulatory climate. Neither Western Union nor the Post Office filed tariffs for the service with the FCC, though both agreed that ECOM was essentially the same as the tariffed Mailgram service. Graphnet (a new public facsimile carrier) challenged the service before both the FCC and the PRC. The Justice Department opposed approval of the service because of potential antitrust violations. The executive agencies began further consideration of the question of the Post Office's role in electronic communications. Congressional committees renewed consideration of the role of the Post Office in electronic mail.

In November 1978 the Common Carrier Bureau of the FCC ruled that Western Union could not provide service to the Post Office except under a tariff. Consequently Western Union filed a tariff for ECOM in January 1979 which the Common Carrier Bureau rejected in April. The tariff was declared to be unlawful because of the discriminatory rate structure and because it would prejudge the determination of whether or not ECOM was subject to FCC control. In further consideration of the issue by the FCC, the commission ruled in August 1979 that the entire ECOM service was a communications service subject to the FCC and that the Post Office should seek authorization as a common carrier and file tariffs with the commission. The Post Office responded in October with an appeals court suit asserting that the FCC was attempting to divest the USPS board of governors and the PRC of their exclusive jurisdiction. The FCC's assertion of jurisdiction and the Post Office refusal to submit to that jurisdiction held up the service for the litigation period.[32]

While the FCC was attempting to restrict the Post Office moves, the Post Office was moving to restrict FCC regulated carriers through the private express statutes. The private express statutes provided a monopoly to the Post Office for delivery of mail, but specifically exempted telegrams in order to allow Western Union messengers freedom to deliver messages. Telegrams were not explicitly defined in the statutes but so long as Western Union had a monopoly on telegraph business, there was little question about what was meant by a telegram. The emergence of new value-added carriers such as Graphnet brought into question exactly what a telegram was.

The application of Graphnet to distribute international telegrams to their final destination caused the FCC to undertake an inquiry into the need for continuing Western Union's monopoly of domestic telegraph. Most respondents advocated competition for telegrams and Western Union itself raised little opposition to new competition. The Public Message Service had been a consistent money loser and had declined from 80 percent of Western Union's revenue in 1947 to 10 percent in 1976. With little opposition to the move, the FCC established a policy of free entry into telegraph services in January 1979 and relieved Western Union of the responsibility of maintaining unprofitable offices in order to provide a telegraph network. Regulation of telegraph services was continued in keeping with the decision to regulate all "second level" or value-added carriers, but much greater freedom for entry, exit, and services was generated by the decision.[33]

The policy of free entry for telegraph services and the emergence of new services with final delivery capability, such as Graphnet's Faxgrams, made the telegram exemption in the private express statutes of greater significance than before. Using a broad definition of telegram, most potential electronic mail could be delivered outside of the Post Office monopoly. The Post Office responded by proposing a definition of telegram that would eliminate most of the new or proposed services. At the end of 1978 the USPS filed a proposed definition of telegram for public comment. Telegrams would be "messages on paper which result from electronic transmissions by carriers which accept the messages at their public offices, or via the public telephone network, in written or oral form and manually enter the messages by alpha or numeric characters into a transmitter for transmission to their public offices in other places."[34] The key restrictions in the definition were that the messages be accepted at the carriers' public offices in written or oral form and that the messages be manually transformed into alpha or numeric characters. The definition would allow the traditional Western Union public message service but would eliminate the non–Post Office delivery of messages originated by computer (as in the pending ECOM proposal) or transmitted by facsimile (as in the Graphnet network). In response to inquiries of the Postal Rate Commission, the USPS denied that the PRC had jurisdiction over the private express statutes.

The PRC listed ECOM as coming under the private express statutes in a Federal Register notice, and said that if it approved ECOM, it "could severely limit such services presently provided by such companies as Xerox, Western Union, and RCA Global Communications."[35] In critical comments on the proposed definition, AT&T disputed the authority of the USPS to determine the boundaries of the private express statutes on its own.[36] The Justice Department challenged the Post Office's plan of "simply overriding other regulatory systems" and detailed a number of problems with "a system whereby one of the possible major players itself would adjudicate the lawfulness of various kinds of private competition."[37] The FCC challenged the Post Office's jurisdiction to impose any restrictions on electronic communications and said that the commission had exclusive authority over such matters. The National Telecommunications and Information Administration (Department of Commerce) accepted USPS authority to make regulations interpreting the private express statutes but challenged the propriety of regulations that contradicted FCC policy. It announced that an administration review of the question was under way. Private companies supported the government agencies' opposition with variations on the same arguments.

At the end of 1979 the future market structure for electronic mail remained uncertain. The USPS was in litigation with the FCC over the latter's authority to regulate electronic mail. The right of the USPS to interpret the private express statutes seemed destined for litigation because of strongly conflicting interpretations of that right. ECOM service was not operational a year after the original proposed service date because of regulatory conflicts. If the Post Office succeeds in obtaining a restrictive definition of the telegram exception to its monopoly, it is likely to accomplish the takeover of the telegraph industry that it missed in 1845. However, a more likely result is additional competition between the Post Office and the telecommunications firms.

Evaluation of Changing Industry Boundaries

The combination of the reduction in regulatory barriers to entry discussed in Chapters 8 and 9 and new entry through technological progress in satellites and computers caused a significant reduction in AT&T's market power. Both events were necessary for substantial new competition. A reduction in regulatory barriers alone left AT&T with many technological and systems advantages over potential entrants. New technologies could not be utilized by potential entrants if multiyear regulatory battles for authorization were necessary. The existence of reduced regulatory barriers for specialized users of old technologies allowed potential entrants to take advantage of new technologies when they became available.

At the time of the original microwave proceedings after World War II, the simultaneous availability of a new technology and a new source of demand created the possibility for new competition. The FCC's restrictive licensing procedure delayed microwave entrants for twenty years by which time AT&T was thoroughly established in the technology. At the time of the domestic satellite authorization, a similar case of a simultaneous new technology and a new source of demand existed. The relaxation of restrictions on cable television created extensive new demand for program transmission to serve the cable television companies. In the satellite case, the combinations of the existing authorizations for microwave competition and political pressure for more competition led to an open-entry policy. New firms established operational experience with satellites at the same time as AT&T and were consequently in a position to expand their operations as the satellite technology improved.

Improvements in computer technology further increased the significance of the satellite decision by expanding the scope of satellite operations. The existence of a communications medium that could reach all over the country in a single transmission allowed the design of a communications network using digital technology that would be completely independent of the AT&T network. Although digital networks were possible (and have been built) using microwave, microwave provided less competitive flexibility than satellite because communications remained distance and location sensitive.

The decreasing cost of satellite communications largely eliminated the technological "natural monopoly" characteristics of the long-distance telecommunications system. Microwave had begun that process by dispensing with the need for physical right of way to lay cables, but competition remained confined to high-density routes because of economies of scale. Because the communications volume on low-density routes was less than the capacity of a single microwave system, an established carrier could exclude entry even with prices above costs. The fixed investment in microwave towers and buildings made it uneconomical to develop temporary microwave lines to an area in which prices were above costs but likely to come down with competition. The advent of larger satellites and portable earth stations will greatly reduce market power even where there is only one existing carrier. With appropriate frequency authorizations, a portable earth station could be set up quickly to provide competition to a monopolist charging above costs. The earth station could then be moved to another location if and when prices were reduced. The existence of potential rapid entry in response to price incentives in any geographical location would prevent even a single firm from having substantial market power.

The rapidly changing technology has caused the traditional regulatory distinctions and industry boundaries to be inconsistent with the actual services available. The attempt to divide between communications and computer services is similar to the 1920 patent cartel's attempt to divide up radio after the development of broadcasting. Either case could be settled by litigation under existing contracts and regulatory decisions, but a legalistic determination ignores the characteristics of the new technology. A series of legal decisions carving up the broadly defined communications market between the Post Office, AT&T, and the computer firms is unlikely to allow free development of innovative services. Combined computer communications services are already hampered by the regulatory lines that prevent firms on either side of the regulatory barrier from making full use of the technological opportunities. The computer inquiry final decision provides hopeful signs of a move toward deregulation and freedom to develop innovative services without concern for traditional industry boundaries, but it faces stringent legal tests prior to implementation.

11 Public Policy Considerations

Congressional Reaction to Competition

As competition increased, AT&T began considering taking its case for monopoly to the public. It had lost several battles in the FCC and was hopeful that a public relations and congressional campaign could restore its protected position. One of the first steps was a trial in Illinois of subscriber reactions to various Bell positions to see which statements would "sell." The results of the trial were not entirely encouraging. The company found the public generally skeptical of Bell's assertions that competition would be detrimental and that Bell had more concern for the consumer than the FCC. The results were summarized as follows:

> The strongest impression one comes away with after reading the research report is a discomforting confirmation of the immensity and complexity of the task facing Bell System information managers . . . the action we protest against (e.g., certifications; proposals to open our markets to competition) are seen as reasonable solutions to problems or as ultimately beneficial to the manager.
>
> If the public is to adopt our position fully it must, in some instances, suspend its convictions and accept the antitheses of some long-held beliefs (e.g., that "free enterprise" means higher prices and lower product or service performance; that the government, in the body of the F.C.C., would deliberately take actions harmful to the public interest; or that modern technology can not overcome the seemingly simple problems presented by interconnection) . . .
>
> The Illinois research argues strongly for a cautious approach to

> public communications, fully examining their potential impact on the audiences involved in advance less we reinforce undesired beliefs or irritate rather than persuade. And, it confirms the need to identify target audiences not only for their potential influence but also in terms of their specific belief structures.[1]

Despite the difficulties of putting the case across to the public, AT&T decided to try. In a speech to the operating company presidents in 1972, AT&T chairman John DeButts suggested that in response to equipment registration proposals, the corporation would take the position that considerable time should be taken in reaching a decision in order to resolve the "far-reaching implications" of terminal competition. The delay was to be used to develop a public relations campaign "so that the long-term public interest implications are clear" and for "putting our own house in order" to meet competition if it came.[2] The public barrage began the following year with a speech by Chairman DeButts before the National Association of Regulatory Utility Commissioners (NARUC) in which he requested a moratorium on competitive experiments until the public interest questions had been settled. This was followed by an internal program to involve many employees in the campaign for legal protection. In outlining the program to Bell managers, considerable emphasis was put on the need for a long-term campaign of widely disseminated information. Some excerpts of the speech outlining the campaign include:

> How to move the public from complete indifference to the concerned action that only the pursuit of self-interest can engender is our joint task . . .
>
> What we are mounting, at least in academic terms, is a challenge to the concept of pure competition and its entire intellectual rationale . . .
>
> It will be won or lost, I'm convinced, by our proving to American consumers that they—and they alone—are going to foot the bill for certification; that they and they alone will suffer the consequences of the slow but deliberate sabotage of the common carrier concept . . .
>
> This will mean a public affairs and a public relations effort of unmatched sophistication and intelligence. . . .
>
> One of the hardest aspects of the job will be to maintain the interest and activity of our people over a prolonged period of time. The battle lines may be drawn, but this will not be but a one-week blitz. Rather it will be a long and hard-fought campaign . . .
>
> It is at the grass-roots that the political process begins—and this is

> where we need your help most . . . in municipal offices, in state houses, in governors' mansions . . . at meetings of civic clubs, engineering societies, service organizations . . . in newspaper editorials, local TV talk shows . . . seeking our allies, probing and maneuvering for consensus.[3]

The public relations campaign on the evils of competition provided a foundation for seeking protective legislation from Congress. The telephone companies drafted a protective bill and began intensive lobbying to gain support for its passage. After very careful preparation, the Consumer Communications Reform Act of 1976 was submitted to Congress at the request of the telephone companies. The bill was cosponsored by 175 members of the House and sixteen senators at the time of its introduction.

The bill and resulting hearings were given wide publicity including mailings to customers and stockholders by the telephone companies. For example, a General Telephone insert to telephone bills stated:

> Since the late 1960's the Federal Communications Commission has introduced what it calls "competition" in providing private, or dedicated communications lines to business users between cities, and terminal equipment—the telephones, switchboards, answering devices and other equipment—in homes and offices.
>
> Each time the telephone industry loses business to a specialized common carrier or terminal equipment manufacturer, it loses the contribution that long distance or specialized services makes toward local service. Eventually, residential and small business customers will bear the burden of the increases in charges for local telephone service.
>
> Proposed legislation, called the Consumer Communications Act of 1976, has been introduced in both houses of Congress to reaffirm the intent of the Communications Act of 1934.
>
> You will be hearing more about this Act in the months to come because of its importance to you and the telephone industry.[4]

An insert in telephone bills for New Jersey Bell stated:

> A few years ago the Federal Communications Commission decided that certain parts of the telephone industry ought to have competition. Other suppliers were encouraged to pick and choose among the telephone industry's profitable services and customers.

As a result, revenues that phone companies traditionally use to keep basic rates low are being lost. If continued, the trend will force a repricing of telephone services and most customers will be faced with sharp increases in rates for basic service. Furthermore, it will be difficult to maintain service levels at their present high standards.

Some members of Congress are concerned about the impact of these departures from national policy. They have introduced legislation in Congress, S. 3129, to reaffirm traditional national communications policy. New Jersey Bell, some 1,600 independent telephone companies, labor unions and others support this legislation.

We think it is time for the public, through its elected representatives, to review this important issue. We are making our views known. Since the issue involves many economic implications for our customers, we encourage you to take a stand too.[5]

The actual bill was formulated as a "reaffirmation" of the intent of Congress in the 1934 act rather than as an entirely new bill. The bill was designed to eliminate the emerging competition in private line services and terminals. With regard to private lines, the bill explicitly stated that "the authorization of lines, facilities, or services of specialized carriers which duplicate the lines, facilities, or services of other telecommunications carriers . . . is . . . contrary to the public interest."[6] In considering any authorization for construction, extension, or renewal of specialized facilities the commission would have to hold an evidentiary hearing and establish that no increased charges would result, that the proposed services were not similar to any services provided by a telephone or telegraph carrier, and that the proposed services could not be provided by a telephone or telegraph common carrier with the burden of proof on the specialized carrier to make the showing.

With regard to terminal equipment, the bill provided that exclusive jurisdiction over terminal equipment would rest with the respective state commissions rather than with the FCC. The provision would eliminate the proposed nationwide registration program and require terminal equipment manufacturers to seek permission from each state commission for attachment. The state commissions had shown less interest in competition than the FCC and were considered more likely to uphold telephone company restrictions.

Despite the impressive number of initial sponsors and intensive telephone industry support, the bill was not enacted. By the time of its introduction, strong competitive forces existed that could be organized against a bill to eliminate them from the market. A broad coalition, including groups with such disparate orientations as the Justice Department, the Computer and Business Equipment Manufacturers Association, the Public Interest Research Group, and the Federal Communications Commission, as well as the specialized carriers and terminal equipment manufacturers, testified against the bill. Richard B. Long, president of the North American Telephone Association (terminal equipment manufacturers), expressed the attitude of the competitors against whom the bill was directed:

> I would like to characterize the current proposal to destroy competition in the telecommunications field as a fraud against the American consumer. American Telephone and Telegraph, which is leading the battle for absolute monopoly, has shown in the past that the needs of the consumer are at best a secondary consideration. This has never been more true than in the proposal to destroy the interconnect industry which through innovation and lower prices has started a revolution in telephony.[7]

After extensive hearings, a much different bill was introduced by Representatives Van Deerlin and Frey (chairman and ranking minority member of the House communications subcommittee which had been considering the Bell bill). The 1978 Van Deerlin-Frey bill had as its basic policy statement "regulation of interstate and foreign telecommunications is necessary, to the extent marketplace forces are deficient."[8] It eliminated most regulation for competitive services. Facilities could be built with notification to the regulatory commission and rates would be presumed to be equitable for all competitive services. Noncompetitive services would still come under rate regulation but tight limitations on the length of hearings were proposed. If a final decision was not reached in nine months, the rate would be presumed to be equitable. The AT&T Consent Decree was abrogated by an express provision allowing any carrier to enter another business through a separate subsidiary. The existing separations formulas were to be replaced by an access fund administered by the regulatory agency and paid into by interstate service carriers. The fund was designated for subsidizing local carriers. Most state regulation was to be eliminated through a provision that companies accepting money from the access fund would come under federal regulation. Monopoly carriers were prohibited from owning a manufacturing subsidiary, a measure designed to force the divestiture of Western Electric by AT&T and of Automatic Electric by GTE.

The Van Deerlin-Frey bill marked the end of the chance for a Bell monopoly bill in Congress. The commitment of the leading members of the House communications subcommittee to a program of increased competition and decreased regulation made it unlikely that serious consideration would be given to a revival of protective legislation. In hearings on the Van Deerlin bill, established monopoly carriers led by AT&T were generally critical and competitors were generally favorable, although all asked for some revisions. AT&T and GTE particularly challenged the substitution of a market standard for standard regulatory language in the bill. There was no provision that service must satisfy the "public interest, convenience, and necessity" or that rates should be "just and reasonable." AT&T also attacked the divestiture requirement. Vice-Chairman Ellinghaus stated:

> To put the matter bluntly, it is my conviction that no single action this subcommittee might undertake would more adversely affect the pace of innovation in telecommunications, the quality of the services the consumer enjoys, or the cost of providing them than the fragmentation of the service process that section 333 would to all intents and purposes accomplish . . .
>
> What is critical to future progress is that there continues to be—at the core of the information infrastructure of this nation—a basic telecommunications network, designed and managed as one entity.[9]

Actual and potential competitors of the telephone companies generally supported the approach of the bill but requested additional protection from tactics of the monopoly carriers. They were concerned that the free-entry provisions of the bill could work against them by extending AT&T's competitive abilities.

The Van Deerlin bill was reintroduced, with significant revisions, in the spring of 1979, along with two Senate bills (by Hollings and Goldwater) to revise the communications regulatory structure. The revised Van Deerlin bill substituted arms-length dealing between AT&T and Western Electric for divestiture. It also proposed total deregulation of intercity communications after ten years and reserved intraexchange voice service regulation to the states. The Hollings and Goldwater bills also were designed to provide greater scope for market forces, but retained more of the traditional regulatory framework than did the Van

Deerlin bill. The Hollings and Goldwater bills provided for the FCC to classify market segments by their degree of competition and provide corresponding amounts of regulation, with full regulation applied to monopoly segments and full deregulation applied to highly competitive segments. All three 1979 bills provided for a phased introduction of the new scheme to avoid excessive disruption to established practices. Senator Hollings described his bill as a handicapping scheme to introduce competition to the industry. He stated:

> An intelligent, flexible handicapping system with built-in mechanisms for tightening or loosening the restraints as competition flourishes or flounders in each market is precisely what the telecommunications industry most needs during the next one or two decades of transition to what we hope will be full competition.[10]

The new bills, particularly the Hollings version, received greater support from AT&T than the original Van Deerlin bill, but the company continued to assert the need for unified management and control of the national network. In hearings and public speeches, AT&T challenged the concept that competition would provide good performance in basic services, while accepting the possibility of competition in terminal equipment and specialized services. AT&T chairman Brown summarized the company's concern that "none of the bills now before Congress seems to us to give adequate attention to assuring the unitary management required not only to design and build a balanced network matched to the needs of a dynamic and ever-changing nation, but to reconfigure that network day by day, hour by hour, minute to minute to match the ebb and flow of communications across the country."[11] Brown stated that Bell approval of new legislation would depend on its meeting four criteria: (1) ability to manage core network, (2) preservation of "the intricate web of working relationships that our business has developed over the course of a hundred years," (3) transition protection of revenue requirements, and (4) protection of "proven organizational arrangements developed over a hundred years" from "hazard before the courts."[12]

The four conditions were essentially impossible to fulfill in any legislation with a deregulation emphasis. In seeking protection of the basic monopoly structure from either competitors or the Justice Department Antitrust Division, the 1979 compromise requirements were similar to the original AT&T goals in the Consumer Communications Reform Act. The chances for passage of a reform bill were further reduced in August when AT&T submitted proposed revisions to the bill which appeared unacceptable to other parties. The revisions suggested AT&T was determined to oppose the basic thrust of the pending bills and passage was considered unlikely with AT&T opposition. In September, Senator Hollings and Representative Van Deerlin both challenged the opposing industry factions to come to some agreement on the outstanding issues rather than reiterating previous positions. Both indicated that the legislation could not go forward without some semblance of industry consensus support.[13]

At the end of 1979 the prospect for major changes in the Communications Act of 1934 in the near future was dim. Widespread dissatisfaction with the current situation existed, but the status quo is a relatively stable arrangement. The failure of the original Bell bill indicated congressional dissatisfaction with overriding the interests of the new competitors in order to restore a monopoly to the telephone carriers. AT&T is unlikely to accept any bill that reduces its current level of regulatory protection. The fact that strongly expressed opposition from part of the industry can block legislation while consensus is necessary for passage makes passage unlikely. Nor is it feasible to gain industry support by cartelization of the industry and protection from new competitors because of the strong interests of potential new competitors (IBM and Xerox, for example) who have already made investments in developing new telecommunications services. It appears that many industry participants prefer to work under the gradual deregulation of the FCC within the existing and familiar legal framework than to accept a new law that does not clearly fit their own interests.

Antitrust Litigation

The beginning of competition in terminal equipment and private-line services renewed the question of the applicability of antitrust standards to AT&T and other telephone companies. The question had been at issue but not fully settled in the 1949–1956 Justice Department suit. The resulting consent decree could have been interpreted to imply antitrust immunity for regulated activities, but there was no explicit recognition of antitrust immunity in the decree. Cases from other industries recognized a potential conflict between antitrust and regulatory standards but did not precisely delineate the relationship between the two standards.[14] Complicating this situation was the fact that some antitrust authority was vested in the regulatory agencies and some explicit exemptions from antitrust were written into regulatory law. The lack of explicit law on the applicability of antitrust standards to AT&T's conduct and the conflicting inferences that could be drawn from existing cases left AT&T's responsibility to satisfy antitrust standards an open question.

So long as AT&T was a monopolist, little incentive existed for applying the antitrust laws against the company. The Justice Department could not reasonably renew its case without some change in circumstances from those existing at the time of the 1956 Consent Decree and there were no private competitors to bring cases. However, with the coming of competition, actual and potential competitors emerged. Frustrated by AT&T's actions to protect its market control and by the FCC's inability or unwillingness to limit AT&T responses, many of the new competitors filed antitrust suits. Between the Carterphone decision and early 1974, thirty-five private antitrust suits were filed against AT&T or its subsidiaries. In late 1974 the Justice Department filed a new suit against AT&T incorporating many of the same claims as the various private suits. After some settlements and several new filings, forty-nine suits were pending against the company in 1979.[15]

Incentive for Bell's competitors to seek redress under the antitrust laws was provided by the success of ITT's suit against General Telephone (GTE), the second largest telephone holding company. GTE maintained a vertically integrated structure through its ownership of telephone operating companies and its acquisition of three equipment subsidiaries (Automatic Electric, Lenkurt Company, and Leich Electric) in the 1950s. When GTE purchased additional telephone operating companies in the 1960s, ITT brought suit, asserting that the vertical structure of GTE and the new acquisitions restrained trade in the equipment industry and prohibited ITT from selling to the newly acquired GTE companies. GTE argued that regulation exempted it from antitrust restrictions. Judge Martin Pence of the Hawaii district court ruled that ITT had standing to sue and GTE was subject to the provisions of the Clayton Act. He furthermore found after trial that GTE's acquisitions were in violation of the Clayton Act and ordered divestiture of GTE's manufacturing subsidiaries. This left the anomalous situation of the vertical integration of GTE being an illegal foreclosure of the equipment market while the far larger vertically integrated AT&T was protected by the 1956 Consent Decree. On appeal, Judge Pence's 1972 decision was set aside and the case was remanded to the Hawaiian court for further proceedings. In 1978 the Hawaiian court again ruled in favor of ITT. Rather than pursue the litigation, the companies then arranged an out-of-court settlement in which GTE agreed to set up an independent equipment evaluation procedure, purchase equipment on an "arm's length basis" from its own subsidiaries and outside suppliers such as ITT, and offer patent licenses to unaffiliated carriers.[16]

In the AT&T suits, the primary issue to date has been the applicability of the antitrust laws to a regulated company. Several of the suits have been dismissed for lack of jurisdiction because of a ruling that AT&T had "implied" antitrust immunity. Others have been allowed to go forward with pretrial preparations. At the moment, there are a number of apparently conflicting lower court rulings with no clear higher court rulings to control the issues. The various courts seem to agree that actions ordered by or explicitly approved by the regulatory agency are exempt from antitrust review, but differ as to the degree of immunity conferred by the general regulatory review process. For example, many of the complaints have centered on AT&T's actions to delay free equipment attachment after the Carterphone decision. Competitors viewed AT&T's requirement for telephone company provided connecting arrangements as an illegal attempt to maintain a monopoly of the terminal market. In the case of *Essential Communications Systems, Inc.* v. *AT&T and New Jersey Bell Telephone Co.,* the district court dismissed the complaint because the tariff was a matter subject to the FCC. In late 1979 the appeals court (third circuit) reversed the dismissal, because the tariff was filed by AT&T and never explicitly approved by the FCC. The court ruled that the fact that the FCC had the power to challenge the tariff but chose to neither approve nor disapprove it did not give antitrust immunity to AT&T.[17]

The rising number of antitrust complaints and competitive situations caused Bell to increase its training operations with regard to competitive selling and antitrust implications. In late 1973 an antitrust seminar was set up for AT&T managers to acquaint them with antitrust concerns of the company. By July 1974, 59,000 persons had been trained in the antitrust seminar.[18] One aspect of the course was training on the problems that careless statements found in employee files during the discovery stages of an antitrust case could cause. In one script, an antitrust trial between Bell and an interconnect company is depicted. An important negative factor in the simulated trial is a memo from a Bell employee expressing his interpretation of reports on competitive equipment in broad terms. The lesson to be drawn from the story was:

> When Simpson wrote that the Engineering letter was "widely interpreted" to mean thus and so, he wasn't telling the truth. He was giving his opinion, he wasn't saying what the collective understanding of the telephone companies was . . . Yet it's terribly damaging . . . The judge and the jury may forget that Simpson didn't know what he was talking about. They may reread that memo and may come to the conclusion that that really was the company policy . . .
>
> And another regrettable thing is this record, this inaccurate record, this record that wasn't even intended to become a company record, need never have been retained. Even if it was a company record it could have been thrown out in six months. The FCC rules didn't require us to retain them any longer than that, there were no outstanding court orders. Instead, this memorandum gets buried in the file until two sharp eyes with contact lenses spotted it.[19]

When questioned at a congressional hearing if the purpose of the script was to encourage employees to destroy incriminating documents, the Bell witness replied: "Well, I think it is a normal management caution on the amount of paper and information that you keep in files. And I would like to say that in our business we are in the telephone business, not the letter-writing business. So we would encourage people to use the telephone."[20]

The 1974 Justice Department suit was based on AT&T conduct in the late 1960s and early 1970s, but it sought many of the same goals as the 1949 suit. Justice asked for a dissolution of AT&T with separate companies for Western Electric (equipment manufacturing), Long Lines (long-distance services), and operating companies (local service). AT&T was accused of monopolizing three markets (equipment, long-distance service, and local service) and using its monopoly power in each market to strengthen its power in the other markets. AT&T responded that it was immune to antitrust prosecution because of regulation, and that its actions had been taken in accord with a regulatory scheme inconsistent with the antitrust laws in order to further the public interest.

The first four years after the Justice case was filed were occupied with legal maneuvering over jurisdiction and discovery procedures. After the assignment of a new judge to the case in 1978, serious preparations for trial began. In September 1978 Judge Harold Greene ruled that his court had jurisdiction for the case and no parts of it were exclusively matters for the FCC. He ordered a strict timetable for discovery and exchange of stipulations and statements of contentions and proof with the goal of a 1980 trial date.[21]

Greene also ruled that the government had a right to see AT&T documents previously offered to private litigants MCI and Litton and determined to be relevant by them. In the private cases, AT&T had produced twelve million pages of documents and the attorneys had selected 2.5 million pages as potentially relevant. Justice requested only the 2.5 million pages and AT&T asserted that it would be unfair to provide them without mixing them back into the 9.5 million rejected pages in order to prevent Justice from taking advantage of private litigants' work. AT&T appealed the order to turn over the selected set rather than the full set of documents as so unfair that it "raises the possibility that a retrial may ultimately be required in this massive lawsuit." The appeals court was unimpressed and the Supreme Court declined review allowing the document production to go forward.[22] The litigation over the rather straightforward document issue and attempt to prevent the government from benefiting from the work of private plaintiffs were part of a defense strategy to overwhelm the government with AT&T's superior resources. In early 1979 AT&T was reported to have assigned three thousand full-time employees to its defense of the case while the Justice Department had assigned twenty-four. AT&T Chairman Brown announced in November 1979 that the company would spend $100 million on its defense of the case in 1979.[23]

In its first statement of contentions in preparation for the trial (January 1979), AT&T continued to rely heavily on its status as a regulated carrier. The company asserted that regulation prevented a finding of monopoly power as a matter of law and that new entry prevented a finding of monopoly power as a matter of fact. AT&T asserted that regulation was responsible for the structure of the industry, that non–cost-related prices were an explicit result of regulatory policies, and that most of the challenged conduct occurred in a period of "uncertainty" when the FCC was departing from long-established policies without explicitly abandoning them. Bell declared that the proposed restructuring was "preposterous" and added:

> Even if some aspect of Bell's conduct were held to violate the antitrust laws, the suggestion that an enterprise as important to the public interest as the Bell System should be fragmented and destroyed by reason of some mistake that was committed during a period of controversy and uncertainty such as that which followed the FCC's change of regulatory policy in 1968 is incredible . . .
>
> There is and can be no justification for such wanton destruction of a valuable national asset, and for that reason alone, there is and can be no justification for the remedy sought by the government here.[24]

While placing great emphasis on the unique problems of antitrust in a regulated industry context, AT&T also pursued standard defensive tactics such as extremely broad views of the relevant market in order to reduce market shares. In its statement on markets, AT&T denied the relevance of market shares but also asserted that the market should be "a single service market consisting of the processing and transmission of information." The equipment market should be "an international market and include computer and data processing equipment as well as traditional telecommunications equipment."[25]

The Justice Department view of the case was that regulation was largely irrelevant. AT&T monopolized in violation of the Sherman Act by the traditional tests: a high market share and specific actions designed to acquire or maintain that market share. The primary market of interest was the overall telecommunications market in the United States of which AT&T had 92.4 percent of the $38.7 billion revenues in 1977. A variety of submarkets were also alleged.[26] The existence of predatory pricing (low rates for competitive private-line services) and exclusionary actions in terminal equipment (restrictive tariffs and connecting arrangements) show Bell's intent to maintain monopoly power.

At the end of 1979 it appeared that at least some of the trials would go forward. AT&T had lost its attempt to get the suits dismissed because of regulatory immunity. However, the key issue is likely to remain the significance of regulation for interpreting AT&T's market share and actions. Disregarding the regulatory context, plaintiff's victories would be practically a certainty. An overwhelming share of the market and acts designed to protect that market traditionally add up to monopolization. On the other hand, if the structure of the industry and actions of AT&T are taken to be largely a result of regulatory policies, AT&T victories are highly likely. Between those two extremes are a wide range of possible rulings on the relationship of the structure and conduct in the industry to the requirements of regulation. The novelty of the issues raised and the wide variety of results shown so far in the jurisdictional rulings of district courts makes it unlikely that the outcome of the cases can be known with any confidence until a final Supreme Court review occurs.

Implications and Conclusions

The preceding account of the telecommunications industry over the last 135 years provides a number of implications for economic theory and public policy. The two are closely intertwined. Policy judgments rest on a theoretical base. The fact that the base may be implicit rather than explicit and occasionally not even recognized by the policy maker does not detract from the fact that a conception of how the industry operates and the results that will flow from any given policy lies behind public actions. This section summarizes the implications of the study. No attempt is made to present a detailed recommendation for policy, but the implications of the findings for policy are briefly spelled out.

The first implication is that industry structures change over time in a nonrandom manner. These changes come as a result of actions of the firms and technological forces. An examination of the context of a particular industry allows one to make predictions of the direction the industry is moving. This point is very significant for fashioning public policy. In antitrust and regulatory contexts, the implicit assumption is often that a policy should be defined for the conditions that exist at one given time. Changes may be expected in the industry but they are expected as a result of policy actions. Little attention is given to the "natural" or market forces of change. Yet the market forces of change may be stronger or act more rapidly than policy changes, as well as causing less disruption in firm organization and management. Policy actions are generally spread out over many years for hearings, reconsiderations, implementation delays, and so on, and provide ample grounds for practically indefinite delay if one of the parties finds delay in its interest. The great changes in the computer industry since the beginning of the 1969 government antitrust suit against IBM are a well-known and often-cited example. Aside from the question of speed, the consideration of market changes allows one to evaluate whether or not policy actions are necessary. Taking an action (such as divestiture) is certainly not free. It imposes substantial legal costs on both sides and substantial managerial burdens for implementation. Insofar as industries are moving in the "right" direction, it may be more economical to allow them to do so than to impose significant changes through government action.

The second implication is that technological change must be considered in a broader context than a single industry. Technological change modifies industry boundaries and changes the conditions of entry. It may create a natural monopoly or destroy one. Demand for new products as a result of changes in another industry may induce new entry into the first industry. In the satellite case, the technology was driven by the United States space program, not by commercial communications needs as such. In the electronics case, the digital technology is driven by the computer industry rather than the commercial communications industry. When relevant technology is being created in outside industries and no legal barriers to entry exist, the firms must either adopt the technology or face new entry. The technological changes can change cost ratios by orders of magnitude rather than by the few percentage points that constitute common static barriers to entry.[27]

The third implication is that systems considerations have a vital competitive effect. A firm with monopoly control over part of the system has an incentive to restrict access to its part in order to extend control over other parts. With private systems, such actions may merely switch the focus of competition from the component level to the systems level. With public systems, such tactics can lead to a total monopoly. To have competition in a public system with some parts monopolized, it is necessary that nondiscriminatory access be maintained. If monopoly power can be broken in all parts of the system, then the systems advantage goes away and the formerly dominant firm has an incentive to cooperate with the establishment of standards.

The fourth implication is that regulation is ineffective in controlling the prices for individual services. Regulation can block new entry and can limit the overall rate of return of the carrier, but cannot control the actual management of the firm. The regulators cannot force a firm to introduce a new product or a new technological advance. They cannot effectively control the individual prices for products, particularly if the prices may be "too low" or predatory. The extensive hearings on the private-line tariffs provide convincing evidence that the managers can maintain tariffs indefinitely regardless of the desires of the regulators.

The fifth implication is that the current pattern of subsidy is largely determined by the carriers themselves rather than specific social policy. AT&T has never been ordered to increase long-distance charges in order to subsidize local service rates. The price discrimination features that exist, such as higher rates for business telephones than residence telephones, were implemented during a time of pure monopoly as part of the monopoly practice prior to any regulation. The actual pattern of subsidies is currently hotly disputed. The conventional wisdom is that businesses subsidize residences, long-distance subsidizes local service, and terminal equipment subsidizes basic service. However, only the long-distance subsidy of local service is clearly established through the separations process. Studies have questioned the existence of a subsidy from terminal equipment (or if it was even paying its own way) and the subsidy of residences by business (because of the heavier average use of a business line than a residence line). AT&T's accounting system does not provide the cost of individual services and special studies vary widely depending on the assumptions made and the purpose of the study. The income distribution implications of the current pattern are uncertain. Subsidies set by the carriers either because of ignorance of actual costs or to achieve certain corporate goals do not necessarily correspond to any subsidies that would be chosen by the public as a whole for the public good. Once any particular pattern of subsidy has been established, it is possible to find justifications for it.

In considering the public policy implications of the study, the first question is the direction market forces are pushing the industry. Telecommunications is a rapidly growing industry, and expected to remain so. Predictions for future growth have been in the 15 to 25 percent range.[28] Consequently, extensive new building of plant will be necessary in order to provide the necessary capacity. Technological change is expected to continue, leading to extensive use of inexpensive digital circuits and switches and many new applications for telecommunications. The interaction of telecommunications products with those in the computer industry is expected to increase. All of these forces are procompetitive and will move the industry toward a competitive structure in the absence of government intervention. Rapid growth attracts new entry and prevents new entrants from merely taking business away from existing capital investment. Technological change attracts new firms by driving prices down and providing an opportunity for innovative companies to enter with a technological advance. The merging of industries provides competition from established firms in the other industries.

The industry is unlikely to ever become perfectly competitive regardless of government policy. The economies of scale and systems effects do not lend themselves to very large numbers of companies. Yet that does not imply that the industry is a natural monopoly. Even a dominant firm industry can achieve performance quite close to that of competition if there are many small firms able to compete within small segments of the industry. It is not necessary for the small firms to have a large market share, either singly or in total. The important fact is that there be relatively free entry into all segments of the industry, and that no segment be totally blocked. The current roadblock to competitive conditions is local distribution. It appears infeasible to have competitive companies stringing wires to individual houses. But while complete local distribution currently appears to have blockaded entry from technological forces, it seems likely that some competition will be feasible in the near future. Xerox's XTEN plan for local distribution among businesses provides one possibility and clearly is competitive with existing wire local distribution schemes. Because competition already exists in long-distance service and in terminal equipment and is possible in local service, no part of the telecommunications network can be predicted to be a clear natural monopoly ten years from now.

The absence of natural monopoly in the industry raises a question of the benefit of continued regulation. Regulation has two major functions. The first is the protection of existing companies from certain kinds of competition in order to protect their revenues and rate relationships. The second is to protect consumers from the monopoly power of the existing firms. To some extent, the second is created by the first. The monopoly power is increased by regulatory protection and then requires regulation to protect the consumer.

The first question in evaluating the wisdom of deregulation is the social desirability of discriminatory pricing and subsidies. The use of subsidies requires continued regulation and entry control. The fundamental market mechanism is for firms to enter industries or products where the prices are above costs in order to make a profit, thus adding to supply and bringing price down to the level of costs. The only way in which a subsidy scheme can operate is by preventing entry into the monopoly goods that provide the subsidy. Alternatively entry can be allowed and a subsidy requirement imposed on the new entrants, as has been done in the case of the specialized carriers' entry into long-distance switched service. If the set of products providing the subsidy is broad or unknown, and the exact amount of the subsidy is unknown, then the practice of using one product to subsidize another becomes an argument for legal barriers to entry and monopolization of the industry.

Three kinds of costs can be identified from a subsidy scheme. The first is a reduction in ordinary allocative efficiency. With subsidies, goods are not allocated in the way that customers would choose to allocate them. Some people who value basic telephone service less than the cost (but more than the subsidized price) will get telephone service. Some people who value long-distance service more than the cost (but less than the price) will not make the long-distance calls. In both cases, resources are not allocated in the best way possible because the non–cost-based prices send out the wrong consumer signals.

The second cost is in information about true costs and available technologies. Because subsidization requires a scheme of monopolization, there is no outside check on managerial mistakes. Future costs are inherently uncertain and any management will make substantial mistakes in estimating them. Under competition, a management who mistakenly believes that a particular technology is uneconomical will be corrected by observing the successful use of that technology by a competitor. In some cases the competitor is a former employee whose ideas were turned down by top management. Such was the case, for example, in the founding of Amdahl Corporation by Gene Amdahl after leaving IBM. Amdahl's decision to leave IBM in 1970 after IBM management turned down his recommendations for large-scale computers resulted in a very successful introduction by Amdahl Corporation of the technology proposed to IBM.[29] Without the ability of a new competitor to enter with ideas passed over by established management, there is no check on the accuracy of established management's expectations. Thus a legal barrier to entry in order to provide excess profits in some services and subsidies in others induces great dependence on the managerial foresight and capabilities of the monopoly firm.

The third cost is reduced responsiveness to customer desires. A company that misjudges customer desires in a competitive market soon learns of its mistake from the success of its competitors. A company that misjudges consumer desires in a monopoly market has no check on its judgment so long as profits continue to be made. There is less actual information (that is, no competitors to watch) and less incentive to gather accurate information about consumer desires in a legally protected monopoly market. The monopolist subject to rate of return regulation can continue to offer traditional products and make the allowable rate of return even if new products would be worthwhile. For example, color telephones were not introduced until the 1950s, and a wide variety of designer telephones were not available from AT&T until after competition began in terminal equipment. The rapid consumer acceptance of designer telephones even at substantially higher prices than ordinary telephones and the simple technological nature of providing colored plastic skins for telephone components suggest that AT&T underestimated consumer demand for variety in telephone styles prior to the competitive era. Although some might argue that there is little social gain from having a Mickey Mouse–shaped telephone instead of an ordinary black one, the fact that consumers have less voice in design decisions under monopoly than under competition is a real cost of a subsidization scheme.

The basic advantage of a subsidization scheme is that it provides more people with basic telephone service. An economic advantage arises from this by extending the range of calling. However, this advantage should not be overestimated. The telephone companies do not necessarily provide service to out-of-the-way places without an extra fee for stringing wire there. Neither Bell nor the major independent companies offered service to many rural areas. Service has been provided to them through an explicit subsidy (primarily low-cost loans) of the Rural Electrification Administration. Thus the subsidy at issue is primarily one of marginal changes in the cost of local telephone service. It is doubtful that a great number of consumers who currently have telephone service would give it up with the price increases occasioned by the end of the subsidy.

The second potential benefit is an income distribution advantage from the subsidization scheme. This is a speculative benefit of very uncertain magnitude because of the absence of information about the real pattern of subsidies and about the pattern of cost transmission from businesses to consumers. Insofar as the number of long distance calls placed per local telephone line rises with the income of the subscriber, the subsidy works to reduce income inequality. While that proposition may be true in general, long-distance calls hardly fall into the class of services purchased only by the very wealthy. It is possible that even many low-income people who use long-distance service sparingly at present would prefer a more cost-based system in which they paid more for local service and less for long-distance calls.

The costs of a subsidy scheme outweigh its benefits. It is a very inefficient system for providing phones to marginal customers who see a low value for them and for providing a potential minor income redistribution function. A government policy of universal telephone service, if it is to exist, should be formulated and administered by government bodies, not by the telephone companies. The preferable way would be direct assistance to low-income consumers for telephone service, such as is now provided for health care, food, housing, and other basic needs. It should be financed out of general tax revenue, not out of the telephone business. If for some reason there needs to be a policy of subsidizing local service with long-distance revenues, it would be more desirable to levy a tax on long-distance service and distribute it to local companies rather than use the current separations process. In that case, there would be no need for monopoly control, although the allocative inefficiencies would still remain.

The second potential reason to retain regulation is to limit the monopoly power of the telephone companies. The current economic power of the established telephone companies is very great even without regulatory protection. The immediate abolition of all regulation would leave consumers exposed to monopoly prices by the carriers. However, the regulatory commissions are not the only protectors of consumers. Antitrust laws also have an important effect on how monopoly power can be used, particularly for so dominant a firm as AT&T. The abolition of regulation, and the consequent abolition of any implied antitrust immunity, would put severe constraints on AT&T's future conduct in order to avoid treble damage suits under the antitrust laws. From the competitors' point of view, such protection would probably be stronger than what they currently gain from the FCC. From the consumers' point of view, it would not. The antitrust conduct prohibitions are designed to prohibit anticompetitive conduct, but not pure high prices. Thus simply raising prices would not be grounds for antitrust action, but it would bring in many competitors, particularly in terminal equipment and long-distance services. If actions were then taken to prevent effective competition by the new entrants, grounds for an antitrust suit would ensue. Consequently, control of market power is not a sufficient reason for continued regulation in the terminal and long-distance markets.

The foregoing considerations lead to a recommendation that long-distance services and terminal equipment should be completely deregulated. Regulation should be retained for local service but be limited to setting maximum prices. No attempt should be made to retain existing entry restrictions or non-cost-based price structures.

Because significant competition in local distribution appears infeasible in the near future, AT&T would retain substantial monopoly power in that segment of the market. Safeguards would be necessary to prevent the use of local distribution power to monopolize the potentially competitive markets. The first possible safeguard is dissolution along the general lines suggested by the Justice Department. Such a dissolution would largely eliminate the extension of monopoly power from local service to other segments and would be desirable if it could be accomplished without cost. However, it is likely to require significant costs in time, litigation expense, and managerial reorganization. The actual costs which would be incurred by dissolution are unknown but are potentially great enough to make dissolution an unwise action.

A second possibility for safeguards consists of simple enforceable conduct restrictions. Such restrictions would include: (1) no personal or use discrimination, and (2) no interconnection restrictions for all equipment which meets appropriate technical standards. The prohibition on personal or use discrimination would eliminate all rates or services based upon the identity of the user. The rate structure for a local service line would be the same whether the line connected a residence, a single business line, a PBX, or a long-distance company to the central office. The rate structure could include a fixed cost and a usage charge and consequently have greater total charges for a heavily used business line than a lightly used residence line, but the rate structure could not vary with the class of customer. Nor could customers be denied service because of their intended use of the line. The prohibition on personal discrimination should include the elimination of all resale restrictions.

There would be no distinction between lines connecting a long-distance company to the local central office and lines connecting a final user. A user would place a long-distance call by calling the long-distance company of his choice. The call would go via the local central office to the long-distance company's office where it would be transmitted over that company's long-distance network to the receiving city and then as a local call from the receiving long-distance office to the final destination. The local telephone company would not distinguish a long-distance call from a local call and the long-distance companies would bill the customer directly. A long-distance competitor could establish service by constructing new facilities or by leasing lines from other carriers for resale. A long-distance company might lease private lines from other carriers and use them to offer dial up service, or construct facilities in certain areas and lease others to complete a network.

The elimination of tying arrangements or interconnection restrictions is necessary in order to allow effective competition in terminals and long distance while monopoly remains in local service. Interconnection eliminates the ability of the monopolist to tie potentially competitive products to its monopoly service.

Both conditions are mild restrictions that are imposed automatically by a competitive market. A competitive company has no incentive to refuse service to anyone willing to pay the going price of the service. It has no ability to tie one product to the purchase of another. Both restrictions can also be imposed by the antitrust laws on dominant firms. Tying is prohibited to firms with even small amounts of market power. Discrimination among buyers is not necessarily a violation of antitrust but can be interpreted as either price discrimination or an attempt to monopolize. Significant personal discrimination by a dominant firm is unlikely to survive antitrust attack. Despite the limited nature of the proposed restrictions, they would make major changes in the business practices of the telephone companies and lead to greatly increased competition.

Under current conditions, deregulation of all long-distance service would leave users in sparsely populated areas subject to unregulated monopoly power because the actual and proposed long-distance competitors cover only the large cities and dense communications routes. However, monopoly power in rural areas could be eliminated by the reallocation of unused radio frequencies to telephone communications. In general, the less densely populated areas also have the least use of radio frequencies. The vast amount of spectrum allocated to UHF television service and underutilized in most areas of the country provides one source of available telephone frequencies. Radio costs are dropping rapidly, allowing radio to provide inexpensive communication where adequate frequencies are available. The ubiquitous citizen's band radio provides one example of a relatively inexpensive radio communications device. With feasible frequency allocations that would not displace any existing services, rural areas could be provided with competitive radio telephone concentrators that would collect the transmissions from a wide area for relay on standard microwave or satellite facilities. It is likely that with appropriate frequency allocations, not only long-distance but local service as well could become competitive in lightly populated areas.

The changes in the telecommunications industry are not all history. They are ongoing and accelerating. The industry is not dying but is advancing and becoming intertwined with other industries. All evidence points to continued rapid development and change in the industry. Technological progress in electronics and transmission technology should bring telecommunications costs down far below today's level. Rising wages and fuel costs should raise the price of postal service and physical transportation. The increased importance of services in general and information in particular relative to manufacturing in an advanced economy will also raise the demand for telecommunications. Public policies for the industry should be designed to fit a rapidly changing, growing, progressive industry with large numbers of potential entrants from diverse technological specialties. No single company is likely to effectively use all the opportunities in telecommunications services in the decades ahead. Nor is any regulatory body likely to be aware of all the possibilities. The only way to ensure full benefits from this industry is to allow wide participation, with the opportunity for many companies to try out their innovations and succeed or fail according to the merit of the plan.

Notes

1. Introduction

1. "Behind AT&T's Change at the Top," *Business Week,* November 6, 1978, 114–135.

2. Ibid., 118; "The New Telephone Industry," *Business Week,* February 13, 1978, 68–78; Bro Uttal, "Selling is No Longer Mickey Mouse at AT&T.," *Fortune,* July 17, 1978, 98–104.

3. See, for example, Joe S. Bain, *Industrial Organization* (New York: John Wiley & Sons, 1959).

4. George Stigler, "The Theory of Economic Regulation," *The Bell Journal of Economics and Management Science,* 2 (Spring 1971), 3–21.

5. F. M. Scherer, *Industrial Market Structure and Economic Performance,* 2nd ed. (Chicago: Rand McNally, 1980), chap. 1; William G. Shepherd, *The Economics of Industrial Organization* (Englewood Cliffs, N.J.: Prentice-Hall, 1979), chap. 1; Alexis P. Jacquemin and Henry W. de Jong, *European Industrial Organization* (New York: John Wiley & Sons, 1977), chap. 1.

6. For a more extensive review of regulatory theory, see Paul L. Joskow and Roger G. Noll, *Regulation in Theory and Practice: An Overview,* California Institute of Technology Social Science Working Paper 213, May 1978.

7. Stigler, "The Theory of Economic Regulation."

8. Donald J. Dewey, "Regulatory Reform," in *Regulation in Further Perspective: The Little Engine That Might,* ed. William G. Shepherd and Thomas G. Gies (Cambridge, Mass.: Ballinger, 1974); Bruce M. Owen and Ronald Braeutigam, *The Regulation Game: Strategic Use of the Administrative Process* (Cambridge, Mass.: Ballinger, 1978), chap. 1.

9. John R. Baldwin, *The Regulatory Agency and the Public Corporation: The Canadian Air Transport Industry* (Cambridge, Mass.: Ballinger, 1975).

10. Victor P. Goldberg, "Regulation and Administered Contracts," *The Bell Journal of Economics,* 7 (Autumn 1976), 426–448.

11. Albert Hirschman, *Exit, Voice, and Loyalty* (Cambridge, Mass.: Harvard University Press, 1970).

12. Gerald Brock, *The U.S. Computer Industry: A Study of Market Power* (Cambridge, Mass.: Ballinger, 1975), chap. 8.

13. *International Business Machines Corp.* v. *United States,* 298 U.S. 131 (1936).

14. Phillip Areeda, *Antitrust Analysis: Problems, Text, Cases,* 2nd ed. (Boston: Little, Brown, 1974), 254–259.

2. The Theory of Dynamic Industry Structures

1. Joe S. Bain, *Industrial Organization* (New York: John Wiley & Sons, 1959); R. E. Caves and M. E. Porter, "From Entry Barriers to Mobility Barriers: Conjectural Decisions and Contrived Deterrence to New Competition," *Quarterly Journal of Economics,* 91 (May 1977), 242–261; A. M. Spence, "Investment Strategy and Growth in a New Market," *The Bell Journal of Economics,* 10 (Spring 1979), 1–19; A. M. Spence, "Entry, Capacity, Investment and Oligopolistic Pricing," *The Bell Journal of Economics,* 8 (Autumn 1977), 534–544; Oliver E. Williamson, *Markets and Hierarchies: Analysis and Antitrust Implications* (New York: Macmillan, 1975), chap. 11.

2. Alexis P. Jacquemin and Henry W. de Jong, *European Industrial Organization* (New York: John Wiley & Sons, 1977); Phillip Areeda, *Antitrust Analysis: Problems, Text, Cases,* 2nd ed. (Boston: Little, Brown, 1974).

3. Spence, "Entry, Capacity, Investment and Oligopolistic Pricing"; Joe S. Bain, *Barriers to New Competition* (Cambridge, Mass.: Harvard University Press, 1956), chap. 4.

4. Darius Gaskins, "Dynamic Limit Pricing: Optimal Pricing under Threat of Entry," *Journal of Economic Theory,* 3 (September 1971), 306–322.

5. Bain, *Barriers to New Competition,* chap. 1; F. Modigliani, "New Developments on the Oligopoly Front," *Journal of Political Economy,* 66 (June 1958), 215–232.

6. Surveys of the technological progress literature are contained in M. I. Kamien and N. L. Schwartz, "Market Structure and Innovation: A Survey," *Journal of Economic Literature,* 13 (March 1975), 1–37; F. M. Scherer, *Industrial Market Structure and Economic Performance,* 2nd ed. (Chicago: Rand McNally, 1980), chaps. 15, 16; William G. Shepherd, *The Economics of Industrial Organization* (Englewood Cliffs, N.J.: Prentice-Hall, 1979), chap. 21; Williamson, *Markets and Hierarchies,* chap. 10.

7. Joseph A. Schumpeter, *Capitalism, Socialism and Democracy,* 3rd ed. (New York: Harper Torchbooks, Harper and Row, 1950), 84–85.

8. Almarin Phillips, *Technology and Market Structure: A Study of the Aircraft Industry* (Lexington, Mass.: Lexington Books, 1971).

3. The Telegraph in the United States

1. Robert L. Thompson, *Wiring a Continent* (Princeton, N.J.: Princeton University Press, 1947), 6–34.

2. Quoted ibid., 32–33.

3. A flavor of the level of sophistication of the early lines is given by this first-person account of the initial Philadelphia–New Jersey line from James D. Reid, *The Telegraph in America* (New York, 1879), 118–119: "One morning, not long after the line was opened, we guessed by the tug of the magnet that the wire was broken not very far away. We had no repairer at Philadelphia, so it became Zook's duty to hunt the break and repair it. I therefore directed him to go to Norristown by cars and walk back till the break was found. He was then to send me a message by planting a piece of wire in the ground and tapping on it with the line wire. That done, he was to find a puddle of water, in which he was to stand, and putting the line wire to his tongue, receive my acknowledgment . . . when, an hour or more afterward, Sam came to the office covered with mud, and madness in his eye, we learned our first lesson in the dangers of line testing and repairs."

4. Quoted in Thompson, *Wiring a Continent,* 85.

5. R. E. Caves and M. E. Porter, "From Entry Barriers to Mobility Barriers: Conjectural Decisions and Contrived Deterrence to New Competition," *Quarterly Journal of Economics,* 91 (May 1977), 242–261.

6. Quoted in Thompson, *Wiring a Continent,* 149.

7. Eyewitness James Reid described the situation as follows in *The Telegraph in America,* 201: "To add to the general unhappiness, the Mississippi burst through the frail levees on the Louisiana coast, washed away the line, and inundated New Orleans. In May, 1850, the office in that city was only accessible by boats. Business of course was suspended. The indebtedness enlarged daily and had reached the amount of $90,000. At the same time an enormous crevasse burst through the levees at Bonnet Carre above New Orleans, and the mad waters of the Mississippi roared through to Lake Ponchartrain. Creditors also, by a kind of instinctive sympathy with nature, were already revenging themselves by cutting down the line. The negro teamsters made their camp fires of the poles and carried off the wire. At some offices the sheriff held the key."

8. Thompson, *Wiring a Continent,* 168–174.

9. Ibid., 275–276, agreement printed on pp. 482–491.

10. Quoted ibid., 280–281.

11. Reid, *The Telegraph in America,* 210.

12. Ibid., 210.

13. Thompson, *Wiring a Continent,* 296–316, agreement reprinted on pp. 504–513.

14. Ibid., 370.

15. Horace Coon, *American Tel & Tel: The Story of a Great Monopoly* (Freeport, N.Y.: Books for Libraries Press, 1939; reprinted 1971), 31.

16. Thompson, *Wiring a Continent,* 399–419; Reid, *The Telegraph in America,* 525.

17. Thompson, *Wiring a Continent,* 424–442.

18. Reid, *The Telegraph in America,* 575.

19. For a discussion of excess capacity as a barrier to entry, see A. Michael Spence, "Entry, Capacity, Investment and Oligopolistic Pricing," *The Bell Journal of Economics,* 8 (Autumn 1977), 534–544.

20. Reid, *The Telegraph in America,* 590–595.

21. James M. Herring and Gerald C. Gross, *Telecommunications: Economics and Regulation* (New York: McGraw-Hill, 1936; reprinted Arno Press, 1974), 3.

22. John Moody, *The Truth about the Trusts* (New York: Moody, 1904); Herring and Gross, *Telecommunications,* 565–587.

23. Lester G. Telser, "Cutthroat Competition and the Long Purse," *Journal of Law and Economics,* 9 (October 1966), 259–277.

24. Reid, *The Telegraph in America,* 564–565, 657–661; Coon, *American Tel & Tel,* 20.

25. Coon, *American Tel & Tel,* 37.

4. The Telephone in the United States

1. Horace Coon, *American Tel & Tel: The Story of A Great Monopoly* (Freeport, N.Y.: Books for Libraries Press, 1939; reprinted 1971), 20; James D. Reid, *The Telegraph in America: Its Founders, Promoters, and Noted Men* (New York, 1879), 657–661.

2. Federal Communications Commission, *Investigation of the Telephone Industry in the United States* (Washington: U.S. Government Printing Office, 1939; reprinted Arno Press, 1974), 26; Reid, *The Telegraph in America,* 643–644; Coon, *American Tel & Tel,* 50.

3. *The Deposition of Alexander Graham Bell,* 1892, reprinted as *The Bell Telephone* (New York: Arno Press, 1974); Coon, *American Tel & Tel,* chap. 4; John Brooks, *Telephone: The First Hundred Years* (New York: Harper & Row, 1975), chap. 2; Joseph C. Goulden, *Monopoly* (New York: G. P. Putnam's Sons, 1968), chap. 3; Frederick Leland Rhodes, *Beginnings of Telephony* (New York: Harper and Brothers Publishers, 1929).

4. Patent reprinted in *The Bell Telephone,* 453–460.

5. Patent reprinted ibid., 463–469.

6. Correspondence between Gray and Bell reprinted ibid.

7. Advertisement reprinted in J. Warren Stehman, *The Financial History of the American Telephone and Telegraph Company* (Boston: Houghton Mifflin, 1925), 6–7; Coon, *American Tel & Tel,* 33; FCC, *Investigation of the Telephone Industry,* 84.

8. Reid, *The Telegraph in America,* 629–632; Coon, *American Tel & Tel,* 37–40.

9. FCC, *Investigation of the Telephone Industry,* 84; Coon, *American Tel & Tel,* 34–35; Brooks, *Telephone,* 65.

10. N. R. Danielian, *A.T.&T.: The Story of Industrial Conquest* (New York: The Vanguard Press, 1939; reprinted Arno Press, 1974), 40–41; Brooks, *Telephone,* 68–69.

11. Goulden, *Monopoly,* 36; Brooks, *Telephone,* 69–70: Stehman, *Financial History of A.T.&T.,* 14–15.

12. Danielian, *A.T.&T.,* 42–43; Stehman, *Financial History of A.T.&T.,* 15; FCC, *Investigation of the Telephone Industry,* 124.

13. Danielian, *A.T.&T.,* 41; FCC, *Investigation of the Telephone Industry,* 124–125.

14. Brooks, *Telephone,* 72; Stehman, *Financial History of A.T.&T.,* 19, also reports the $1,000 quotation but questions whether any transactions took place at that price.

15. Stehman, *Financial History of A.T.&T.,* 17.

16. Address of Theodore Vail to the National Geographic Society, March 7, 1916, A.T.&T. Historical File, New York.

17. W. S. Comanor, "Vertical Mergers, Market Power, and the Antitrust Laws," *American Economic Review,* 57 (May 1967), 254–265; Frederick R. Warren-Boulton, *Vertical Control of Markets: Business and Labor Practices* (Cambridge, Mass.: Ballinger, 1978).

18. Stehman, *Financial History of A.T.&T.,* 20–26.

19. Quoted in Danielian, *A.T.&T.,* 43.

20. Vail notes, undated, A.T.&T. Historical File, New York.

21. Vail speech, "Lest We Forget," undated, A.T.&T. Historical File, New York.

22. Quoted in FCC, *Investigation of the Telephone Industry,* 215–216.

23. Rhodes, *Beginnings of Telephony* contains an extensive account of the litigation.

24. Vail, "Lest We Forget."

25. FCC, *Investigation of the Telephone Industry,* 7, 353.

26. Vail, "Lest We Forget."

27. FCC, *Investigation of the Telephone Industry,* 29–30.

28. Ibid., 35; Stehman, *Financial History of A.T.&T.,* 26–27.

29. FCC, *Investigation of the Telephone Industry,* 129–136; Stehman, *Financial History of A.T.&T.,* 18.

30. FCC, *Investigation of the Telephone Industry,* 126; Richard Gabel, "The Early Competitive Era in Telephone Communication, 1893–1920," *Law and Contemporary Problems,* 34 (1969), 352; Jeffrey G. Williamson, *Late Nineteenth-Century American Development: A General Equilibrium History* (London: Cambridge University Press, 1974), 97.

31. John Moody, *The Truth about the Trusts* (New York: Moody, 1904), 381–384; FCC, *Investigation of the Telephone Industry,* 56.

32. Initial unsuccessful attempts to lay an Atlantic cable led Western Union to believe that the only way to connect the United States and Europe

was via the Pacific. A Western Union subsidiary had constructed a telegraph line up the Canadian coast and part of the way across Siberia when it received word of the successful transatlantic cable and abandoned the project. See Robert L. Thompson, *Wiring a Continent: The History of the Telegraph Industry in the United States 1832–1866* (Princeton, N.J.: Princeton University Press, 1947), 434.

33. James M. Herring and Gerald C. Gross, *Telecommunications: Economics and Regulation* (New York: McGraw-Hill, 1936; reprinted Arno Press, 1974), 1–25; Moody, *The Truth about the Trusts,* 381.

34. Harry B. MacMeal, *The Story of Independent Telephony* (Chicago: Independent Pioneer Telephone Association, 1934), 26–27.

35. Ibid., 33–35.

36. FCC, *Investigation of the Telephone Industry,* 126–137.

37. Stehman, *Financial History of A.T.&T.,* 80–91.

38. Ibid., 85–95.

39. MacMeal, *Independent Telephony,* 85; Stehman, *Financial History of A.T.&T.,* 97–100; Danielian, *A.T.&T.,* 48; FCC, *Investigation of the Telephone Industry,* 129.

40. Quoted in Danielian, *A.T.& T.,* 97.

41. Quoted ibid., 98.

42. Ibid., 97; MacMeal, *Independent Telephony,* 85; FCC, *Investigation of the Telephone Industry,* 134.

43. Stehman, *Financial History of A.T.&T.,* 26–27; FCC, *Investigation of the Telephone Industry,* 135.

44. FCC, *Investigation of the Telephone Industry,* 504; Gabel, "Early Competitive Era in Telephone Communication," 352.

45. Quoted in FCC, *Investigation of the Telephone Industry,* 352.

46. Ibid., 222–223, 353.

47. Ibid., 131; Danielian, *A.T.&T.,* 48.

48. FCC, *Investigation of the Telephone Industry,* 132.

49. Ibid., 132; Danielian, *A.T.&T.,* 49–50; David Hemenway, *Prices and Choices: Microeconomic Vignettes* (Cambridge, Mass.: Ballinger, 1977), chap. 16.

50. MacMeal, *Independent Telephony,* 141.

51. Ibid., 138–147; FCC, *Investigation of the Telephone Industry,* 137.

52. FCC, *Investigation of the Telephone Industry,* 137.

53. Computed from figures ibid., 128–129.

54. Ibid., 129–136.

5. The Telephone and Telegraph in Europe

1. Jeffrey Kieve, *The Electric Telegraph: A Social and Economic History* (Newton Abbot, Eng.: David & Charles, 1973), 18–23; Robert L. Thompson, *Wiring a Continent: The History of the Telegraph Industry in the United States 1832–1866* (Princeton, N.J.: Princeton University Press, 1947), 14–26.

2. Kieve, *The Electric Telegraph,* 29–51; Thompson, *Wiring a Continent,* 241.

3. Kieve, *The Electric Telegraph,* 83–92.

4. Ibid., 53–91.

5. Ibid., chaps. 6–8.

6. British figures from ibid., 183; American figures from John Moody, *The Truth about the Trusts* (New York: Moody, 1904), 384.

7. Kieve, *The Electric Telegraph,* 204–206.

8. Ibid., 195; A. R. Bennett, *The Telephone Systems of the Continent of Europe* (London, 1895; reprinted Arno Press, 1974), 1–31.

9. Kieve, *The Electric Telegraph,* chap. 11; Bennett, *Telephone Systems of Europe,* 1–31.

10. Kieve, *The Electric Telegraph,* 195; Moody, *The Truth about the Trusts,* 384.

11. A. N. Holcombe, *Public Ownership of Telephones on the Continent of Europe* (Cambridge, Mass.: Harvard University Press, 1911), 3–5.

12. Ibid., 270–271.

13. Ibid., 278.

14. Some of the early French problems remain today. The *Wall Street Journal* of January 24, 1979, reported that there was a waiting list of 1.5 million persons for telephone service in France. The average waiting time for service is 8.5 months. Corporations make an interest-free loan of up to $3,400 for each phone line ordered.

15. Holcombe, *Public Ownership of Telephones,* 280, 302–303.

16. Ibid., 335.

17. Ibid., 10–53, 127–182.

18. Bennett, *Telephone Systems of Europe,* 334.

19. Holcombe, *Public Ownership of Telephones,* 386–387.

20. Computed from figures ibid., 418–438, and Federal Communications Commission, *Investigation of the Telephone Industry in the United States* (Washington, D.C.: U.S. Government Printing Office, 1939; reprinted Arno Press, 1974), 129.

6. The Decline of Competition, 1907–1935

1. Harvey Leibenstein, *Beyond Economic Man: A New Foundation for Microeconomics* (Cambridge, Mass.: Harvard University Press, 1976); Oliver E. Williamson, *Markets and Hierarchies: Analysis and Antitrust Implications* (New York: Macmillan, 1975).

2. Federal Communications Commission, *Investigation of the Telephone Industry in the United States* (Washington, D.C.: U.S. Government Printing Office, 1939; reprinted Arno Press, 1974), 135.

3. N. R. Danielian, *A.T.&T.: The Story of Industrial Conquest* (New York: Vanguard Press, 1939; reprinted Arno Press, 1974), 50.

4. Quoted in FCC, *Investigation of the Telephone Industry,* 90.

5. Danielian, *A.T.&T.*, 61–66.
6. Quoted ibid., 56–57.
7. Ibid., 66–75.
8. Harry B. MacMeal, *The Story of Independent Telephony* (Chicago: Independent Pioneer Telephone Association, 1934), 172.
9. Ibid., 177.
10. Ibid., 179.
11. Ibid., 186.
12. FCC, *Investigation of the Telephone Industry,* 128–136.
13. MacMeal, *Independent Telephony,* 181–190.
14. *Standard Oil Co.* v. *United States,* 221 U.S. 1 (1911); *United States* v. *American Tobacco Co.,* 221 U.S. 106 (1911).
15. FCC, *Investigation of the Telephone Industry,* 139–140; Danielian, *A.T. & T.*, 75–77; Kingsbury Commitment letter reprinted in MacMeal, *Independent Telephony,* 204–207.
16. FCC, *Investigation of the Telephone Industry,* 139–143.
17. Ibid., 139–143.
18. J. Warren Stehman, *The Financial History of the American Telephone and Telegraph Company* (Boston: Houghton Mifflin, 1925), 149.
19. Ibid., 164–167.
20. Paul W. MacAvoy, *The Economic Effects of Regulation: The Trunkline Railroad Cartels and the Interstate Commerce Commission before 1900* (Cambridge, Mass.: MIT Press, 1965) concluded on p. 201: "The trunk-line railroads benefited from regulation—the rules of the Interstate Commerce Commission against personal and locational discrimination, and against setting secret rates, had made it possible to maintain basic long-distance grain rates of 23 to 25 cents for seven years. It had not been possible to maintain 21–22 cent rates previously, nor was it possible to maintain 17–20 cent rates in the years following the weakening of regulation."
21. American Telephone & Telegraph, *1907 Annual Report* (Boston: Alfred Mudge & Son, 1908), 18.
22. American Telephone & Telegraph, *1910 Annual Report* (Boston: Alfred Mudge & Son, 1911), 33.
23. T. N. Vail, "Some Truths and Some Conclusions," speech to Vermont State Grange, December 14, 1915, A.T.&T. Historical File, New York.
24. Quoted in Gabel, "Early Competitive Era in Telephone Communication," 357.
25. Stehman, *Financial History of A.T.&T.*, 261–262.
26. Paul Schubert, *The Electric Word: The Rise of Radio* (New York: Macmillan, 1928), chaps. 1–4.
27. Ibid., 68, Danielian, *A.T.&T.*, 103.
28. Quoted in Danielian, *A.T.&T.*, 104–105.
29. Quoted in FCC, *Investigation of the Telephone Industry,* 187.
30. Quoted ibid., 192.
31. Quoted ibid., 210.
32. Leonard S. Reich, "Research, Patents, and the Struggle to Control

Radio: A Study of Big Business and the Uses of Industrial Research," *Business History Review,* 51 (1977), 216; Schubert, *The Rise of Radio,* 127–148.

33. Danielian, *A.T.&T.,* 109; Schubert, *The Rise of Radio,* 128–131.

34. Reich, "Research, Patents, and the Struggle to Control Radio," 214; Schubert, *The Rise of Radio,* 147–157.

35. FCC, *Investigation of the Telephone Industry,* 224–225; Schubert, *The Rise of Radio,* 174–177.

36. Reich, "Research, Patents, and the Struggle to Control Radio," 217–218; FCC, *Investigation of the Telephone Industry,* 225–226; Schubert, *The Rise of Radio,* 207.

37. Quoted in FCC, *Investigation of the Telephone Industry,* 209.

38. Schubert, *The Rise of Radio,* 197–200.

39. Reich, "Research, Patents, and the Struggle to Control Radio," 221–225; FCC, *Investigation of the Telephone Industry,* chap. 13.

40. Quoted in FCC, *Investigation of the Telephone Industry,* 389.

41. Ibid., 228–230.

42. Ibid., 230; Danielian, *A.T.&T.,* 134.

43. Quoted in FCC, *Investigation of the Telephone Industry,* 232.

44. Quoted in Danielian, *A.T.&T.,* 143.

45. Ibid., 140–143.

46. Ibid., 147–148; FCC, *Investigation of the Telephone Industry,* 406–409.

47. Quoted in FCC, *Investigation of the Telephone Industry,* 209–210.

48. Ibid., 197, 234.

7. The Era of Regulated Monopoly, 1934–1956

1. James M. Herring and Gerald C. Gross, *Telecommunications: Economics and Regulation* (New York: McGraw-Hill, 1936; reprinted Arno Press, 1974), chaps. 14, 15, original act reprinted in Appendix A.

2. FCC, *Investigation of the Telephone Industry,* 602.

3. Ibid., 602; John Brooks, *Telephone: The First Hundred Years* (New York: Harper & Row, 1975), 196–198.

4. Eugene Lyons, *David Sarnoff: A Biography* (New York: Harper & Row, 1966), 220, 275.

5. Donald C. Beelar, "Cables in the Sky and the Struggle for Their Control," *Federal Communications Bar Journal,* 21 (1967), 27–28.

6. William G. Shepherd, "The Competitive Margin in Communications," in *Technological Change in Regulated Industries,* ed. William M. Capron (Washington, D.C.: The Brookings Institution, 1971), 113–115; F. M. Scherer, "The Development of the TD-X and TD-2 Microwave Relay Systems in Bell Telephone Laboratories" (Case study, Harvard University Graduate School of Business Administration, 1960).

7. Beelar, "Cables in the Sky," 29–32.

8. *Federal Communications Commission Reports,* 42 (1949), 10, with referenced report beginning on p. 1; future references cited in form 42 *F.C.C.* 1,

10 (1949) in which numbers refer to the volume number, beginning page of report, referenced page, and year, respectively.

9. Quoted ibid., 13.

10. Ibid.

11. *Philco Corporation et al.* v. *American Telephone and Telegraph Company,* 80 *F. Supp.* 397 (1948).

12. 42 *F.C.C.* 1, 23 (1949).

13. 17 *F.C.C.* 132 (1952).

14. Beelar, "Cables in the Sky," 33–37.

15. Brooks, *Telephone,* 232.

16. U.S., Congress, House, Committee on the Judiciary, *Hearings before the Antitrust Subcommittee,* 85th Cong., 2nd sess., pt. 2, vol. 1 (March 25–May 22, 1958), *Complaint* reprinted, 1719–1795.

17. A.T.&T., *Answer,* reprinted ibid., 1799–1844.

18. Brooks, *Telephone,* 234–236.

19. Wilson to Lilienthal, July 1, 1949, reprinted in *Hearings before the Antitrust Subcommittee,* 1713–1714.

20. Robert Lovett to the attorney general, March 20, 1952, reprinted ibid., 1878–1879.

21. C. E. Wilson to H. Brownell Jr., July 10, 1953, reprinted ibid., 2029–2031.

22. T. B. Price, memorandum of March 3, 1954, referring to meeting of June 27, 1953, reprinted ibid., 1953–1955.

23. A.T.&T., "Memorandum for the Attorney General," (April 1953), 38, reprinted ibid., 1958–1995.

24. *Consent Decree,* reprinted ibid., 1845–1861.

25. Quoted ibid., 2640.

26. Sample agreements reprinted ibid., 3261–3344.

8. New Competition I: The Long-Distance Market

1. Leonard Waverman, "The Regulation of Intercity Telecommunications," in *Promoting Competition in Regulated Markets,* ed. Almarin Phillips (Washington, D.C.: The Brookings Institution, 1975), 211.

2. 27 *F.C.C.* 359, 364 (1959).

3. Ibid., 388.

4. Ibid., 377–381.

5. Ibid., 404–405.

6. Ibid., 403–414.

7. *Radio Regulation,* 20 (1960), 1603; future references cited in form 20 *R.R.* 1603 (1960).

8. Computed from figures in 38 *F.C.C.* 370, 385–386 (1964).

9. 38 *F.C.C.* 370, 379 (1964).

10. 37 *F.C.C.* 1111 (1964).

11. 40 *R.R. 2d* 1289 (1977).
12. 60 *F.C.C.. 2d* 261 (1976).
13. 66 *F.C.C. 2d* 132 (1977).
14. Quoted in *Telecommunications Reports,* March 5, 1979, 6.
15. 9 *F.C.C. 2d* 30, 37 (1967).
16. 6 *R.R. 2d* 953 (1966).
17. 16 *R.R. 2d* 1037 (1969).
18. Quoted in 16 *R.R. 2d* 1037, 1067 (1969).
19. 27 *F.C.C. 2d* 380 (1971).
20. "Statement of William G. McGowan," in *Hearings before the Subcommittee on Antitrust and Monopoly of the Committee on the Judiciary,* U.S. Senate, 93rd Cong. on S. 1167, Pt. 2 (July 30–August 2, 1973) (Washington, D.C.: U.S. Government Printing Office, 1973), 716–763.
21. 27 *F.C.C. 2d* 380, 382 (1971).
22. 29 *F.C.C. 2d* 870, 920 (1971).
23. "Transmission Facilities," *Datapro Reports on Data Communications* (Delran, N.J.: Datapro Research Corporation, various dates), vol. 3.
24. 29 *F.C.C. 2d* 870, 940 (1971).
25. 29 *R.R. 2d* 1589, 1617 (1974).
26. 47 *F.C.C. 2d* 660 (1974); 60 *F.C.C. 2d* 939 (1976).
27. 20 *F.C.C. 2d* 383 (1969).
28. Reprinted in *Hearings before the Subcommittee on Antitrust and Monopoly of the Committee on the Judiciary,* U.S. Senate, 93rd Cong. on S. 1167, Pt. 6 (July 9–31, 1974) (Washington, D.C.: U.S. Government Printing Office, 1974), 4640.
29. Ibid., 4644.
30. 9 *F.C.C. 2d* 30 (1967).
31. 58 *F.C.C. 2d* 362 (1976).
32. 59 *F.C.C. 2d* 428 (1976).
33. *Telecommunications Reports,* March 26, 1979, 19–21.
34. Quoted in *Telecommunications Reports,* May 21, 1979, 40.
35. Quoted in *Telecommunications Reports,* September 24, 1979, 11.
36. Ibid.
37. "Transmission Facilities," *Datapro Reports.*
38. FCC 76–879 (First Report, Docket 20,003, September 23, 1976), 59.
39. FCC, *Investigation of the Telephone Industry,* 129.
40. 62 *F.C.C. 2d* 815 (1976); 62 *F.C.C. 2d* 774 (1977).
41. 60 *F.C.C. 2d* 25 (1976).
42. Harry Newton, "Communications Lines," *Business Communications Review,* 7 (September–October 1977), 30–31.
43. "Regulatory Developments: Slip Sliding Away," *Business Communications Review,* 8 (March–April, 1978), 46–47; *Telecommunications Reports* (May 15, 1978), 3–5; *Telecommunications Reports* (December 4, 1978), 8.
44. *Telecommunications Reports,* May 30, 1978, 1–5.
45. *Telecommunications Reports,* June 26, 1978, 6–8.

46. *Telecommunications Reports,* October 30, 1978, 6–9, 23–25; November 6, 1978, 1–4, 24–26; December 11, 1978, 1–2, 24; December 18, 1978, 12–14.

47. *Telecommunications Reports,* August 13, 1979, 8; September 3, 1979, 5.

9. New Competition II: Terminal Equipment

1. Extensive factual information on terminal equipment is contained in the decision of *International Telephone and Telegraph Corporation* v. *General Telephone & Electronics Corporation and Hawaiian Telephone Company* (1972), reprinted in *Hearings before the Subcommittee on Antitrust and Monopoly of the Committee on the Judiciary,* U.S. Senate, 93rd Cong. Pt. 2 (July 30–August 2, 1973) (Washington, D.C.: U.S. Government Printing Office, 1973), 640–704.

2. Eugene Singer, *Antitrust Economics: Selected Legal Cases and Economic Models* (Englewood Cliffs, N.J.: Prentice-Hall, 1968), chaps. 15–17; Phillip Areeda, *Antitrust Analysis: Problems, Text, Cases,* 2nd ed. (Boston: Little, Brown, 1974), 568–634.

3. I once questioned a Bell Laboratories scientist on why the company was so slow to introduce such an apparently simple innovation as the colored telephone. His answer was that company policy required phones to last for forty years and extensive research was necessary to develop dyes that would last that long.

4. *Hush-A-Phone Corporation* v. *U.S. and FCC,* 238 F.2d 266 at 269 (1956).

5. 22 *F.C.C.* 112, 114 (1957).

6. Quoted in 13 *F.C.C. 2d* 420, 437 (1968).

7. Ibid., 434, 435.

8. Ibid.

9. *International Business Machines Corp.* v. *United States,* 298 U.S. 131 (1936).

10. 15 *F.C.C. 2d* 605 (1968).

11. "More Firms Buying Their Own Phone Systems; 'Interconnect' Becomes a $150 Million Business," *Wall Street Journal,* July 30, 1973.

12. "Minutes of A.T.&T. Presidents Conference, November 9–12, 1970," reprinted in *Hearings before the Subcommittee on Antitrust and Monpoly of the Committee on the Judiciary,* U.S. Senate, 93rd Cong., Pt. 6 (July 9–31, 1974) (Washington: U.S. Government Printing Office, 1974), 4618–4619.

13. "Material Prepared for the Use of Southwestern Bell at Bell System Presidents Conference of October 30–November 8, 1971," reprinted ibid., 4627–4628.

14. "Minutes of Presidents Conference, May 8, 1972," reprinted ibid., 4639.

15. Computed from figures in F.C.C. Docket 20,003, Bell Exhibit 20 (April 21, 1975).

16. F.C.C. Docket 20,003, Bell Exhibit 30, Comments of the Computer and Business Equipment Manufacturers Association; FCC 76–879 (First Report, Docket 20,003, September 23, 1976), para. 240–246.

17. F.C.C. Docket 20,003, Bell Exhibit 47, Appendix A.

18. The same tactic was used by IBM in offering price reductions on computer peripheral equipment after the advent of competition in that submarket. Rather than instituting across-the-board price cuts, new equipment (some of which was technologically the same as old equipment) was offered at reduced prices to customers willing to make the effort to remove old equipment and install new. See Gerald Brock, *The U.S. Computer Industry: A Study of Market Power* (Cambridge, Mass.: Ballinger, 1975), chap. 8.

19. 35 *F.C.C. 2d* 539 (1972).

20. 56 *F.C.C. 2d* 593, 601 (1975).

21. Ibid., 600; 59 *F.C.C. 2d* 83 (1976).

22. "Regulatory Developments: Status of Major FCC Inquiries," *Business Communications Review* 8 (January-February 1978), 41; 61 *F.C.C. 2d* 396 (1976); *Telecommunications Reports,* July 17, 1978, 4–6.

23. *Telecommunications Reports,* September 3, 1979, 23.

24. Bro Uttal, "Selling is No Longer Mickey Mouse at A.T.&T.," *Fortune,* July 17, 1978, 98–104.

25. 56 *F.C.C. 2d* 593, 611 (1975).

26. *Telecommunications Reports,* June 5, 1978, 9.

27. Ibid.; "Illegal Telephones are a Growing Problem for Ma Bell's Operating Companies," *Wall Street Journal,* August 23, 1979, 1.

28. 64 *F.C.C. 2d* 1039; *Telecommunications Reports,* May 30, 1978, 22.

10. New Competition III: Changing Industry Boundaries

1. F. M. Scherer, "The Development of the TD-X and TD-2 Microwave Relay Systems in Bell Telephone Laboratories" (Case study, Harvard University Graduate School of Business Administration, 1960), 43.

2. Jonathan F. Galloway, *The Politics and Technology of Satellite Communications* (Lexington, Mass.: Lexington Books, 1972), 28.

3. Ibid., chaps. 3, 4.

4. James Martin, *Future Developments in Telecommunications,* 2nd ed. (Englewood Cliffs, N.J.: Prentice-Hall, 1977), 472; Satellite system applications in FCC Docket 16,495.

5. Peter Flanigan to Dean Burch, January 23, 1970, reprinted in 22 *F.C.C. 2d* 86, 125–128 (1970).

6. 22 *F.C.C. 2d* 86, 93 (1970).

7. 35 *F.C.C. 2d* 844 (1972).

8. 43 *F.C.C. 2d* 1141, 1163–1164 (1973).

9. The Muzak network provides a good example of the tradeoff between satellite capacity utilization and antenna size. By spreading the transmission of four music channels (the equivalent of twenty voice circuits) over a full satellite transponder (1,200 voice circuit capacity), good results could be obtained with a four-foot antenna in an earth station costing under $1,000. See Martin, *Future Developments in Telecommunications,* 486–491.

10. 34 *F.C.C. 2d* 1, 18 (1972).

11. 34 *F.C.C. 2d* 1, 22–23 (1972); "FCC Decides to Stop Requiring Licenses for Antennas Receiving Satellite Signals," *Wall Street Journal,* October 19, 1979; *Telecommunications Reports,* October 22, 1979, 17; (December 11, 1978), 25.

12. "Transmission Facilities," *Datapro Reports on Data Communications* (Delran, N.J.: Datapro Research Corp., various dates), vol. 3; The Center for Communications Management, Inc., *The Guide to Communication Services* (Ramsey, N.J., various dates).

13. "Transmission Facilities," *Datapro Reports.*

14. "RCA Aides' Faces are Red at Show for Cable-TV Trade," *Wall Street Journal,* December 13, 1979; *Telecommunications Reports,* August 13, 1979, 27; July 2, 1979, 28; December 11, 1978, 25.

15. *Telecommunications Reports,* June 26, 1978, 1–4.

16. Martin, *Future Developments in Communications,* 52.

17. 28 *F.C.C. 2d* 267 (1971).

18. *I.C.C.* v. *Delaware L. & W. R.R. Co.* 220 U.S. 235 (1911).

19. 60 *F.C.C. 2d* 261 (1976); *American Telephone and Telegraph* v. *FCC,* 572 *F.2d* 17.

20. F.C.C. 79–307 (July 2, 1979; *Telecommunications Reports,* July 9, 1979, 1–8.

21. *Telecommunications Reports,* April 14, 1980, 1–8, 33–36.

22. *Telecommunications Reports,* May 12, 1980, 4.

23. 62 *F.C.C. 2d* 997 (1977).

24. *Telecommunications Reports,* September 4, 1978, 1–4; October 1, 1979, 11–13.

25. *Telecommunications Reports,* April 30, 1979, 50; (May 5, 1979), 32.

26. Bro Uttal, "I.B.M. Reaches for a Golden Future in the Heavens," *Fortune,* June, 1977, 172–184; Martin, *Future Developments in Telecommunications,* 484–486.

27. *Telecommunications Reports,* November 20, 1978, 1–4.

28. "Facsimile," *Datapro Reports on Data Communications* (Delran, N.J.: Datapro Research Corp., various dates), vol. 3.

29. Ibid.

30. Martin, *Future Developments in Telecommunications,* 285.

31. *Telecommunications Reports,* September 11, 1978, 14.

32. *Telecommunications Reports,* November 20, 1978, 18; April 9, 1979, 1; August 6, 1979, 8; October 22, 1979, 12.

33. *Telecommunications Reports,* January 29, 1979, 1.

34. Quoted in *Telecommunications Reports,* January 29, 1979, 29.

35. Quoted in *Telecommunications Reports,* March 5, 1979, 28.

36. *Telecommunications Reports,* February 19, 1979, 9–10.

37. *Telecommunications Reports,* March 19, 1979, 16–19.

11. Public Policy Considerations

1. "Prices and Services—Subscriber Reactions to Copy Statements (Illinois)," reprinted in *Hearings before the Subcommittee on Antitrust and Monopoly of the Committee on the Judiciary,* U.S. Senate, 93rd Cong. (Washington, D.C.: U.S. Government Printing Office, 1974), pt. 5, pp. 3455–3459.

2. "Closing Remarks of J. D. DeButts to Presidents Conference, May 12, 1972," reprinted ibid., pt. 6, pp. 4641–4646.

3. "Speech of Paul M. Lund on Public Relations," reprinted ibid., pt. 6, pp. 4647–4652.

4. "General Telephone Lines," (October 1976), reprinted in *Hearings before the Subcommittee on Communications of the Committee on Interstate and Foreign Commerce,* House of Representatives, 94th Cong. (Washington, D.C.: U.S. Government Printing Office, 1977), 150–153.

5. Quoted ibid., 36.

6. Bill and analysis reprinted ibid., 763–780.

7. "Statement of Richard B. Long," ibid., 545.

8. House bill 13015, 95th Cong., 2nd sess.; summary and excerpts in *Telecommunications Reports,* June 12, 1978, 1–9, 27–33.

9. Summary and excerpts of testimony in *Telecommunications Reports,* July 31, 1978, 1–8, 25–33.

10. House bill 3333, Senate bills 622 and 611, 96th Cong. 1st sess.; Summaries and analysis in *Telecommunications Reports,* March 19, 1979, 1–10, 37–40; March 26, 1979, 18; April 2, 1979, 1–10.

11. Quoted in *Telecommunications Reports,* June 18, 1979, 8–9.

12. Ibid.

13. *Telecommunications Reports,* August 6, 1979, 1–5, 21–22; September 17, 1979, 1–3.

14. Phillip Areeda, *Antitrust Analysis: Problems, Text, Cases,* 2nd ed. (Boston: Little, Brown, 1974), 105–114.

15. *Hearings before the Subcommittee on Antitrust and Monopoly,* pt. 6, p. 4616; "AT&T Sees Another Record Year in '80; Nine Month Profit Rose More than 6%," *Wall Street Journal,* November 15, 1979, 5.

16. *Telecommunications Reports,* December 26, 1978, 13–14.

17. *Telecommunications Reports,* November 5, 1979, 8–9.

18. "Statement of Frank A. McDermott," in *Hearings before the Subcommittee on Antitrust and Monopoly,* pt. 6, p. 4078.

19. "What Is in Our Files," A.T.&T. Antitrust Review Seminar Script, reprinted ibid., pt. 6, pp. 4190–4196.

20. "Statement of Frank A. McDermott," ibid., pt. 6, p. 4079.

21. *Telecommunications Reports,* September 18, 1978, 3–9.

22. Ibid.; *Telecommunications Reports,* October 30, 1978, 11–12; January 15, 1979, 14.

23. "AT&T Sees Another Record Year," *Wall Street Journal,* November 15, 1979, 5.

24. Quoted in *Telecommunications Reports,* January 15, 1979, 5–8.

25. *Telecommunications Reports,* August 13, 1979, 6–8.

26. *Telecommunications Reports,* September 3, 1979, 7–10.

27. Bain defined "very high" barriers to entry as those which allowed established firms to elevate price 10 percent above minimal cost without attracting entry. That level would not protect a firm against even one year of technological progress in basic electronics components. See Joe S. Bain, *Barriers to New Competition* (Cambridge, Mass.: Harvard University Press, 1956), 170.

28. "The New New Telephone Industry," *Business Week,* February 13, 1978, 68–78.

29. Gene M. Amdahl, "The Early Chapters of the PCM Story," *Datamation,* February 1979, 113–116.

Index